Math Goes to the Movies

MATH GOES TO THE MOVIES

Burkard Polster
Marty Ross

The Johns Hopkins University Press
Baltimore

© 2012 The Johns Hopkins University Press
All rights reserved. Published 2012
Printed in the United States of America on acid-free paper
9 8 7 6 5 4 3 2 1

The Johns Hopkins University Press
2715 North Charles Street
Baltimore, Maryland 21218-4363
www.press.jhu.edu

Library of Congress Control Number: 2011943465

ISBN 13: 978-1-4214-0483-7 (hc)
ISBN 10: 1-4214-0483-4 (hc)
ISBN 13: 978-1-4214-0484-4 (pbk)
ISBN 10: 1-4214-0484-2 (pbk)

A catalog record for this book is available from the British Library.

Special discounts are available for bulk purchases of this book. For more information,
please contact Special Sales at 410-516-6936 or specialsales@press.jhu.edu.

The Johns Hopkins University Press uses environmentally friendly book materials,
including recycled text paper that is composed of at least 30 percent post-consumer
waste, whenever possible.

Contents

Preface   xi

Part I Movies

1   Good Math Hunting   3
   1.1   How to Become a Math Consultant   3
   1.2   Actors Versus Blackboards   4
   1.3   Lambeau, the Fields Medalist   5
   1.4   A Professor Turns Actor   6
   1.5   Mathematics: "Perceval"   7
   1.6   Students in Action   8
   1.7   Mathematics: Graph Theory 1   9
   1.8   Mathematics: Eigenvalues   12
   1.9   Mathematics: Graph Theory 2   13
   1.10   Mathematics: Graph Theory 3   16
   1.11   Mathematics: The Salieri Scene   17
   1.12   Further References and Remarks   18
   1.13   Parodies   18

2   The Clever Hand Behind *A Beautiful Mind*   21
   2.1   Real Math Beats Fake Math   23
   2.2   Signature Scene   23
   2.3   The Riemann Hypothesis   27
   2.4   The Blonde and the Nobel Prize   33
   2.5   Hand Double   36
   2.6   Bits and Pieces   37

3   Escalante Stands and Delivers   41
   3.1   Welcome to the Finger Man   42
   3.2   Filling the Hole   42
   3.3   Let X Be the Number of Girlfriends   45

3.4   Newton Was an Idiot   46
3.5   The Students Stand and Deliver, Again   50
3.6   Will the Real Jaime Escalante Please Stand Up?   51

4   The Annotated Pi Files   **53**
4.1   Max the Mathematician   53
4.2   Mathematics Is the Language of Nature   54
4.3   Pattern in Pi   54
4.4   Numerology: Father + Mother = Child   57
4.5   The Fibonacci Numbers and the Golden Ratio   58
4.6   Archimedes the Goldfish   63
4.7   Coincidence, or Is It?   64
4.8   Back to the Golden Ratio   66
4.9   Staring into the Sun   67
4.10   The Name of God   69
4.11   Living Happily Ever After   69

5   Nitpicking in Mathmagic Land   **71**
5.1   Tiny Nitpick: What Is Pi?   71
5.2   Historical Nitpick: Pythagoras and the Pythagoreans   72
5.3   Small Nitpick: Pythagorean Music   73
5.4   Medium Nitpick: The Golden Section   74
5.5   Large Nitpick: Three-Cushion Billiards   77
5.6   NOTpick   82
5.7   Notes   83

6   Escape from the Cube   **85**
6.1   The Cube in *Cube*   85
6.2   First Insight: The Power of Primes   86
6.3   How to Avoid Prime Numbers   87
6.4   Second Insight: The Cube in Coordinates   89
6.5   Third Insight: Permutations   90
6.6   Final Insight: Prime Powers   95
6.7   Other Cubes   96

7   The Incredible Shrinking Room   **97**
7.1   How Good a Puzzler Are You?   98
7.2   Answers to the Puzzles   99
7.3   Notes   102

8   Murder in the Hot House   **103**
8.1   The Story   103
8.2   Let's Kill Some Mathematicians   104
8.3   The Writing on the Wall   106
8.4   Let's Run Away and Join the Circus   107

9   A Word Problem for Die Hards 109
    9.1   Playing Billiards with the Die Hard Problem 110
    9.2   A Recipe 111
    9.3   Thwarting a Different Simon 112
    9.4   The Least Common Multiple 112
    9.5   References 113

10   7 × 13 = 28 115
    10.1   First Proof: Bogus Division 115
    10.2   Second Proof: Bogus Multiplication 117
    10.3   Third Proof: Bogus Addition 118
    10.4   Play it Again, Abbott 119
    10.5   General Bogus Math 120

11   One Mirror Has Two Faces, Two Mirrors Have ... 121
    11.1   Real Life 121
    11.2   Prime Numbers 122
    11.3   Calculus 123
    11.4   Mathematical Miscellany 127
    11.5   The Other Mirror 130
    11.6   Notes 131

12   It's My Turn for Some Serious Mathematics 133
    12.1   The Snake Rears Its Lovely Head 133
    12.2   The Classification of Finite Simple Groups 136
    12.3   Questions and Answers 139
    12.4   Notes 139

Part II Mathematics

13   Beautiful Math, or Better Off Dead 143
    13.1   The Direct Approach 143
    13.2   The Poetic Approach 146
    13.3   The All-Singing, All-Dancing Approach 148
    13.4   Any Place, Any Time 149

14   Pythagoras and Fermat at the Movies 151
    14.1   Pythagoras's Theorem 151
    14.2   Fermat's Last Theorem 154
    14.3   Fermat's Last Tango 157

15   Survival in the Fourth Dimension 161
    15.1   Time, Space, Both, or What? 162
    15.2   The Hypercube Via Analogy 163

15.3 Picturing the Hypercube   165
15.4 Dimension Drive   171
15.5 Intersections   173
15.6 The Hypersphere   175

16 To Infinity, and Beyond!   177
16.1 Mystical Musings   178
16.2 Toward Infinity, but Getting Lost   178
16.3 Toward Infinity, and Almost Getting There   179
16.4 Toward the Infinitely Small: Romantic Zeno   180
16.5 To Infinity: Are We There Yet?   181
16.6 Fishing with a Really Big Net   182
16.7 Infinity Pays Its Way   184
16.8 Pretty Patterns, Pretty Pi   186
16.9 Golden Infinity   188
16.10 A Golden Argument   189
16.11 Poetic Summation   190

17 Problem Corner   191
17.1 Problems for Wizkids, and a Wizdog   191
17.2 Math Quiz for Mortals   195
17.3 Devilish Problems   197
17.4 Crazy Problems for Extra Credit   198

18 Money-Back Bloopers   199
18.1 Boosting the Computer   199
18.2 Playing the Percentages   199
18.3 The Curse of Pi   201
18.4 Prime Problems   202
18.5 Slips of the Tongue   204
18.6 Less Is More   204
18.7 Simple Arithmetic?   204
18.8 A Very Tough Quadratic   205
18.9 The Algebra Problem   205
18.10 A Tough Competition   205
18.11 Scary Geometry   206

19 The Funny Files   209
19.1 Sex   209
19.2 Geometry   210
19.3 Arithmetic   212
19.4 Algebra and Word Problems   218
19.5 Mathematicians in Action   221
19.6 Doing the Impossible   223

19.7 What Are the Odds?   226
19.8 Odds And Ends   227

Part III Lists

20 People Lists   231
20.1 Real Mathematicians   231
20.2 Female Mathematicians   235
20.3 Interesting Math Teachers and Classroom Scenes   237
20.4 Wizkids   239
20.5 Mathematicians and Murder   240
20.6 Famous Actors Being Mathematical   244
20.7 Math Consultants   250

21 Topics Lists   251
21.1 Counting to 101   251
21.2 Math Titles but No Math   252
21.3 Pythagoras's Theorem and Fermat's Last Theorem   252
21.4 Geometry   254
21.5 Higher Dimensions   258
21.6 Topology   260
21.7 Golden Ratio and Fibonacci Numbers   261
21.8 Pi   262
21.9 Prime Numbers and Number Theory   263
21.10 Chaos, Fractals, and Dynamical Systems   266
21.11 Communicating with Aliens   266
21.12 Code Breaking   267
21.13 Calculus   268
21.14 Infinity   272
21.15 Paradoxes   273
21.16 Probability, Gambling, and Percentages   275
21.17 Famous Formulas, Identities, and Magic Squares   276
21.18 Mathematical Games   279

Movie Index   281

Preface

About ten years ago, on a whim, we began to collect movies containing mathematics. Now, as a consequence of that whim, we own a library of more than 800 movies on DVD, VHS, 16 mm, Laserdisc, and some strange thing called a CED video disc. The movies range from those expressly about mathematicians, to those that, for whatever reason, just happen to have a snippet of humorous mathematical dialogue.

Over the years, we have found that it is not only professional mathematicians who find the fun in this cinematic mathematics. Just about everybody is charmed by Meg Ryan explaining Zeno's paradox in *I.Q.*, Danny Kaye singing about Pythagoras's theorem in *Merry Andrew*, Lou Costello explaining to Bud Abbott why $7 \times 13 = 28$ in *In the Navy*, and so on. Our book, and its accompanying page on our website www.qedcat.com, is an attempt to identify, organize, and engagingly present this fascinating and funny material.

Our intended audience comprises mathematicians, math students, teachers, and more generally anyone who enjoys a Saturday night at the movies (popcorn and all), has some appreciation of mathematics, and has a sense of fun. That is, we're writing for people like ourselves and our friends, raised on a diet of both modern movie-going and mathematical popularizations by the likes of Martin Gardner.

Math in the Movies So Far

There have been a number of previous attempts to organize and popularize movie mathematics. The following are particularly notable:

- A number of excellent websites (all easily found with Google). Most notable are Arnold G. Reinhold's The Math in the Movies page and Alex Kasman's Mathematical Fiction page, both containing lists and summaries of movies with mathematical content. Oliver Knill's website is a great source for mathematical movie clips. There are also specialized websites, which are extremely comprehensive: Andrew Nestler's and

Sarah J. Greenwald's terrific website dedicated to mathematics in *The Simpsons*; the math section of the Spanish website La Indoblable Página de Bender Bending Rodriguez, dedicated to mathematics in *Futurama*; and Wolfram's Numb3rs website, dedicated to math in the TV series *NUMB3RS*.

- A number of mathematical movie festivals, in particular that organized by Michele Emmer and Michele Mulazzani in Bologna in 2000, and the Cinemath festival organized by Robert Osserman and Michael Singer in Berkeley in 2002.
- Several articles about mathematics in the movies by some of the most well known popularizers of mathematics, including Keith Devlin and Ivars Peterson.[1]
- The books and articles by Michele Emmer.[2]
- The very comprehensive Spanish book *Las Matemáticas en el Cine* by Alfonso Jesús Población Sáez (Spanish Mathematical Society, 2006). Alfonso also has a companion *Cine y mathemàticas* website.
- The book *The Numbers Behind NUMB3RS: Solving Crime with Mathematics* by Keith Devlin and Gary Lorden (Plume, 2007).

Our Book

Our goal is to complement and significantly extend the available information about math in the movies. There is no intention to be truly encyclopedic and to document every last occurrence of math in the movies. There are simply too many boring scenes that are better left unmentioned. Also, except for a few compelling references, we have tended to stick to the movies and to leave TV alone. In particular, we have included only a small fraction of the wonderful math in *The Simpsons*, *Futurama*, and *Numb3rs*. These are huge and worthy projects in themselves, and all have been successfully accomplished elsewhere.

We have, however, attempted to be functionally encyclopedic: in conjunction with our website, we have endeavored to hunt down and to describe all the "good stuff," the scenes we believe are of general appeal and usefulness. Furthermore, our emphasis is really on the math and the fun of seeing it on the big screen, not on anything else. The flipside is that our book probably offers little to experts in cinema studies and serious movie critics.

Our source material ranges from the hilariously nonsensical to very nice adaptations of beautiful mathematics. The movies vary from *1984*, containing just one (critical) line of mathematical dialogue, to a movie such as *π*, which has math in almost every scene.

---

[1] Keith Devlin, Math becomes way cool, Devlin's Angle, November 1998 and Ivars Peterson, Abbott and Costello's wacky math, Ivars Peterson's Math Trek, March 2000.

[2] See, in particular, Michele Emmer and Mirella Manaresi, *Mathematics, Art, Technology, and Cinema* (Springer, New York, 2003).

To gain control of all this material, to get a proper sense of "what is there," we began with some serious organization and documentation:

- Systematically excerpting and commenting on relevant dialogues, screen shots, directors' comments on DVDs, published interviews, and so on.
- Reconstructing the content of blackboards, exam papers, and the like, which appear in the various scenes.
- Identifying mistakes and making what sense we could of confusing scenes.
- Reviewing all the documentation and commentary we could find on movie mathematics, including the websites and writings mentioned above.
- Whenever possible, interviewing mathematicians who acted as consultants for the movies.
- Checking out novels and plays, where movies were based on such sources.

When we set out to write this book, we intended to cover everything, and given the amount of material we've looked at we could probably do just that—in twenty thick volumes. We contemplated various ways of managing the material, for instance by restricting ourselves to feature movies. However, doing so would have excluded many highlights.

After experimenting with various formats, we decided on a collage approach. So we worked on creating self-contained chapters, permitting the chapters to differ radically in terms of content, style, and language. This allowed us to style each chapter as seemed best for the specific material.

Some of the chapters are based on a theme (for example, the fourth dimension or infinity); in these chapters we thread the relevant material from different movies into mathematical surveys, with the aim of both informing and entertaining. Other chapters are built around a single very mathematical movie, such as π or *Good Will Hunting*; these chapters are as much stories of the movies as of the specific mathematics. Two final sections contain annotated lists of movies according to their content: real mathematicians, famous actors playing mathematicians, geometry, probability, and so on.

Our Website

Some movie math is really, *really* dull. Even limiting to the good stuff, trying to cram everything into this one book would have resulted in a bloated, boring mess. However, we did want to be comprehensive, and to make available as much of our source material as possible. To that end, on our website www.qedcat.com we have listed each of the 800+ math movies we have seen, together with a capsule summary of each. Our website also includes excerpted dialogues from many of the movies and links to many movie clips. This material is freely available to anyone interested.

As we learn of movies, new and old, we will add details to our website. By doing so, we hope to extend our book and to keep it up to date.

How to Get Copies of the Movies and Clips

This is a book about movies. So, to properly appreciate many parts of the book, it really helps to have seen the movies we're discussing. This is particularly true for the chapters dedicated to individual movies, such as *A Beautiful Mind* and *Good Will Hunting*.

Ideally this book would have been complemented by a DVD containing mathematical snippets from the movies, but it proved impossible to obtain the necessary permissions. Fortunately, with the increasing popularity of DVDs, and with online video sites such as YouTube, many of the best movie math scenes are now very easy to find. To facilitate this, on our website we maintain a list of links to many of the clips.

Acknowledgments

We would like to thank our many friends and colleagues who spotted math in the movies for us.

Many thanks to the math consultants of various movies who generously gave their time to talk or write to us about their experiences: Len Adleman (*Sneakers*), Dave Bayer (*A Beautiful Mind*), Benedict H. Gross (*It's My Turn*), Tom Kuiper and Linda Wald (*Contact*), Patrick O'Donnell (*Good Will Hunting*), Henry C. Pinkham (*The Mirror Has Two Faces*), and David W. Pravica (*Cube*).

We would also like to thank Jean Doyen and Michele Emmer for their very helpful advice; Hendrik van Maldeghem for translating Dutch dialogue from *Antonia's Line*; the Statens Ljud-Och Bildarkiv in Stockholm for making available a copy of *Berget på Månenes Baksida*; David Alciatore for his expert insights on the game of billiards and the diamond system; David Thompson for his thoughts on the origins of the Abbott and Costello arithmetic routine; Vincent Burke, Trevor Lipscombe, and Jennifer Slater, our editors at the Johns Hopkins University Press, for their efforts in seeing this book through to publication; Narayanan Sabapathy for his very careful proofreading of a final draft; and Jeff Ross for his brotherly dedication to critiquing and proofreading.

Finally, our huge thanks to Anu and Ying, for their invaluable advice, infinite patience, and inestimable support.

Part I
Movies

Watch the movies!

Part I is devoted to some of the most famous mathematical movies, such as π and *Good Will Hunting*. Most of these movies are readily available on DVD, and we have written the chapters with the assumption that the reader has watched the movies.

Exceptions are the relatively obscure *It's My Turn* (chapter 12), and the episode "Hot House" of the Australian TV series *City Homicide* (chapter 8). In case you cannot find DVDs of these, our website includes links to clips and transcripts of the major mathematical scenes, and more background information.

Two exceptions of a different kind are the famous jugs problem in *Die Hard: With a Vengeance* (1995) (chapter 9); and the discussion of one of the most famous and funny mathematical movie scenes, in which Lou Costello convinces Bud Abbott, that $7 \times 13 = 28$ (chapter 10). Clips of these scenes are easy to find, and there are also links on our website.

Chapter 1
Good Math Hunting

Will Hunting, played by Matt Damon, is a mathematical wunderkind. Will has no formal mathematical education, and when we first encounter him as a twenty-year-old in *Good Will Hunting* (1997), he is working as a janitor at the Massachusetts Institute of Technology (MIT). Will solves a couple of (supposedly) difficult problems left on a blackboard by the math professor Lambeau (Stellan Starsgård). Lambeau is suitably impressed and tries to help Will sort out his somewhat messed up life. To this end, Lambeau enlists the help of his psychologist colleague, Sean (Robin Williams).

This is not really a movie about mathematics or mathematicians, but there is plenty of math to be spotted. In fact, having talked to Patrick O'Donnell, a professor of physics at the University of Toronto and the man responsible for most of the mathematics in the movie, we think it would have been great to have included a "math consultant's commentary" on the DVD.

One of our goals here is to provide this missing commentary, a fascinating peek behind the scenes of the mathematical making of *Good Will Hunting*. This commentary is based on our telephone interview with Patrick O'Donnell, who turns out to be a raconteur. In order to capture the spirit and the fun of O'Donnell's anecdotes, the following is a fairly literal transcript of the relevant parts of the interview.

A second goal of this chapter is to explain the mathematics behind the problems that Will solves. These problems turn out to be quite accessible to anybody unintimidated by matrices. However, any reader should feel free to skip our forays into "higher" mathematics: none of the mathematics is required to appreciate either the movie or Patrick O'Donnell's engaging commentary.

1.1 How to Become a Math Consultant

O'DONNELL: "First of all, the really funny thing is how I became the consultant. A bunch of graduate students and some of our faculty here went out for lunch near the university. We were sitting in this little Vietnamese restaurant and this man and these two women came in and sat near us. The man kept looking at me and I thought: 'He must know me or something or I know him.'

"Anyway, we left the restaurant and were walking up the street. About twenty feet up the street or so, one of these two women came running up to me and said, 'We're making a movie and the director would like to have you as an extra. Would you agree?' I said, 'Oh sure that sounds good,' and so we went back, they took a photograph, and they asked me what I did.

"About an hour later they turned up at my office. They had tracked me down, it turned out. When they came in, my board was full of jottings and wave functions and things like that, and the director was very impressed by this, I think, and asked me, 'Would you be a consultant?'

"So, the next day Gus Van Sant, the director came back with the script, told me the story, and said he wanted the math to be technically correct but he didn't have to understand it. He didn't want made up things.

"At that stage they had only one math scene written out completely, and that turned out to be totally wrong. It looked like something in number theory, but it had to do with counting patterns and squares of something, and it didn't make sense. So I had to think of something else. However, in the end the replacement we did for that scene didn't get into the movie."

1.2 Actors Versus Blackboards

O'DONNELL: "One thing I found about Hollywood is that none of the actors could write on a blackboard. Not at all [*laughs*]. Essentially all the writing you see, maybe one wasn't quite, is mine. I did some of it where it really was not appropriate because it was too much of mine.

"In fact, we were invited to New York to the premiere. My wife had seen the dailies in the scenes I am the actor in, but had never seen the rest of the movie. She said to me at one point, 'This is terrible, I see your handwriting everywhere but somebody else is finishing it off!'

"There is one scene (based on a visit by the director to MIT) where he saw two postdocs coming out of a room and there's a board in the corridor, and one wrote something and the other wrote something and they didn't speak, you know. They were communicating through the equations, or whatever they were doing. So, the director wanted a scene like that and I said, 'Just do it the way they do it.'

"Well, by the time it was time to shoot that scene, I had realized that nobody could write. So, I had to dumb it down and dumb it down and dumb it down. All they had to do was score out. I think it was something with factors, one over the other. However, they could not do this and act at the same time.

"So, really, it was dumbed down. In fact, I said to the director, 'Well you just do it the way you describe it. Somebody writes on the board, puts down the chalk and the other person picks it up and writes something.' But he said, 'Oh no, no.' He had these two guys sitting down and it became like a dance and it lasted three minutes or so. In fact the continuity girl said at one point, 'I thought this is only a twenty-second scene' [*laughs*]. And it turned out it *was* only a twenty-second scene by the time they finished editing it."

1.3 Lambeau, the Fields Medalist

0:32

A classroom at Bunker Hill Community College, where Sean teaches psychology.

LAMBEAU: Hello, Sean.

SEAN: Hey, Gerry. Um—Ladies and Gentlemen, we're in the presence of greatness. Professor Gerald Lambeau. Fields Medal winner for combinatory mathematics.

LAMBEAU: Hello.

SEAN: Anyone know what the Fields Medal is? It's a really big deal. It's like the Nobel Prize for math, except they only give it out once every four years. It's a great thing. It's an amazing honor.

O'DONNELL: "The other thing was how to treat Lambeau. I mean they were making a big point about him being a Fields medalist. And, in fact, when they first came to me about consulting they had made him a mathematician who had won a Nobel Prize, and you can't do that because Nobel expressly forbade that since his wife or his mistress ran off with a mathematician.[1]

"So Gus said, 'Well, we'll make him a physicist.' Then they went to Boston and met Sheldon Glashow, who is a Nobel Prize winner in physics, and he said, 'No, no you've got to make him a mathematician because you can do this in mathematics like Ramanujan, but you can't have that in physics because you need to know the facts, so to speak, as well.' So they made him back into a Fields medalist at that point. He just took these things on the fly. You know, 'Just do it!'

"The actor who plays Lambeau [*Stellan Starsgård*] worked for many years with Ingmar Bergman. I once asked him, 'How do you find making movies there compared to that.' He said in the movies you can ad lib a lot and extemporize. Well, getting him to change a word was like pulling teeth.

"There was one scene, I call it the Salieri scene ('You are so much greater than me'), and I think in the original script it had something like 'twenty mathematicians in the world can understand this,' which is something that came out of Einstein's days. I remember as a boy my father used to say that only twelve people in the world understood Einstein's theory. I think that

---

[1] This is a famous though apocryphal story, with Nobel's fellow Swede Gösta Mittag-Leffler usually assigned the role of offending mathematician. It is true that there is no Nobel Prize for mathematics, but there is no evidence that the lack of such a prize is due to Nobel having animosity toward Mittag-Leffler, or any mathematician, for any reason.

came out of the 1930s Chicago science fair or something. It is quite a popular thing, people always say that.

"I tried to get him to change it to 'mathematicians know this,' instead of putting a number to count it, 'cause I said there are not many mathematicians in the world anyway, and so you don't have to say twenty [*laughs*]. Well if you look at that scene in the movie you see he actually falters over it—I got him to agree.

"With Robin Williams everything was made up with him—the scene in the diner where they go up to the table and argue, every shot he changed it. In fact, the part where he says something about the Unabomber, that wasn't in the script originally, and then when he started it he started altering it, playing around with it and asking the barman, 'Who is the Unabomber?' or 'Who is Ted Kaczynski?'—or some variation of that each time. He did improvisations on the script most of the time [*laughs*]."

1.4 A Professor Turns Actor

O'DONNELL: "I'm the drunk at the bar [*figure 1.1*]. The scene starts with Robin Williams repeating part of a risqué joke that Matt Damon did earlier on, about a stewardess going for coffee, and I'm yelling at him basically. I'm supposed to be this barfly who is sort of really annoyed with life, and then they walk away from me up to a table. It took twelve hours to do that scene. And the next morning it took about five hours to do the argument of the two of them round the table.

"Once you get over the rush of seeing a movie being made, you just leave. It is very boring, actually. Typically in a scene they did maybe six different views of the scene, plus before they start they do something called blocking out, and that is getting people into position, and so on. And when they start filming each of those ways of looking at the scene they might do as many as a hundred takes. So it really can get very tedious after a while.

"It turns out there was another actor by the name of Patrick O'Donnell who plays Marty the custodian. We'd seen the movie in New York just before Christmas and when I got back it came out in Canada, maybe the first week or second week of January. One of my friends in computer science had gone to see it before we went to see it. And so when I was in New York I really didn't pay attention, it was too exciting to follow that stuff, to be taking it all in. But he called me up and said, 'You had two parts in it.' He said, 'You were also a janitor.' I hadn't seen it really at that point. I said, 'No, no, no,' and he said, 'Thank goodness, I thought makeup was good but not that good' [*laughs*]. Anyway I never met that man."

It is worth noting that our O'Donnell (not the janitor) is a multi-movie actor: he also played a math professor in the short film *Infinitely Near* (1999).

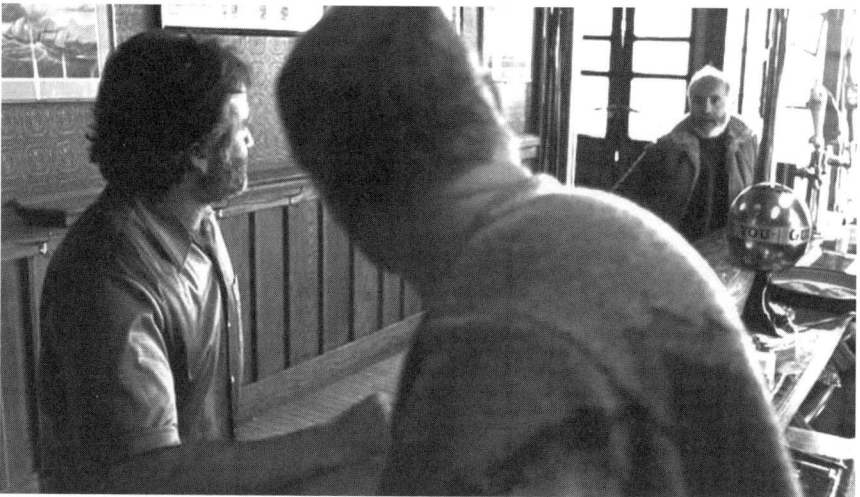

Fig. 1.1 Patrick O'Donnell (facing the camera), making a cameo as a drunk.

1.5 Mathematics: "Perceval"

O'DONNELL: "I have a friend, a chemical engineer. He went to see the movie for two reasons. One was to see me and the other was to see my mistakes. And he said, 'I didn't see you.' He's still convinced that the opening scene is a mistake. Because at least in the text books that I had been looking at at the time, it was clear that people now say integral of f, they don't write $\mathrm{d}x$.[2] We have always used $\mathrm{d}x$ and engineers presumably always use $\mathrm{d}x$. He was complaining that was all wrong because there was no $\mathrm{d}x$ in any of those integrals.

"I think at the end of that scene on the blackboard, the one thing I was worried about in that one—it was shot in the physics department here in a big lecture hall, and it has four blackboards and what I was really worried about was filling all four blackboards, you know having something on them.

"But I think maybe the last equation is integral mod f squared $\mathrm{d}x$ or something. And so what I had to do with Stellan Skarsgård is have him do the last x I think it was, put the period in maybe, but say the words 'integral mod f squared $\mathrm{d}x$" as he kind of did something with the x. So there was a $\mathrm{d}x$ somewhere. Yes definitely there was a $\mathrm{d}x$ in that Parceval theorem. And he mispronounced it and I couldn't get him to fix that one up. Very rigid." [Figure 1.2]

---

[2] O'Donnell is referring to the two common ways of writing an integral: $\int f(x)\,\mathrm{d}x$ and $\int f$. The former notation is common in school and undergraduate classes, and in more applied areas of mathematics, while pure mathematicians tend to use the latter notation.

0:03
Lecture theatre.
LAMBEAU: Mod f squared d*x. So, please finish Perceval.*[3]

Fig. 1.2 Lambeau announces the first superhard problem, and mispronounces "Parseval."

1.6 Students in Action

It is in the Perceval scene that Lambeau announces the first superhard problem: the problem on the hallway blackboard that we shall see in a moment (figure 1.3). Despite what Lambeau says here, this is not "an advanced Fourier system," but is rather a fairly simple problem from algebraic graph theory.

O'DONNELL: "There were many more scenes [*filmed*] than made it into the movie, and I had no input on the editing. There was, in fact, for this scene, following from what he said, a nifty Fourier problem, filmed in a similar manner to what actually appeared in the movie.

"I had given an exam just the Christmas beforehand, which was a Fourier series question, which is one I rather like because nothing seems to be happening in it. Normally when you do these Fourier series, say with a guitar string that is plucked and then let go. That's a typical one, but this one is a model of a piano where the string is at rest and you bang it a very short impulse with a sharp instrument, and that's all you do. The string's at rest,

---

[3] This refers to Parseval's theorem, visible on the bottom of the right blackboard in figure 1.2. This and the other math on the blackboards is part of what is usually referred to as *Fourier analysis*.

Fig. 1.3 In the MIT hallway, Lambeau discovers Will's solution to his problem.

nothing has happened to the string in terms of distortion, and it's totally at rest except for that little bit.

"Anyway I'm writing this up and the professor comes down with a bunch of extras who are students. I'm just about finishing it off and they stop and gather round in a U-shape, and then they have a break and this young guy comes up to me and asks, 'What's that about?' and I say, 'It's a Fourier series' and he asks, 'Do you teach this?' I said, 'I set this as a Christmas exam.' And he said, 'Where?' I said, 'The University of Toronto.' He says, 'I'm in first year in U of T doing physics and math. Would I get a problem like that?' [*laughs*] But that scene didn't get in."

1.7 Mathematics: Graph Theory 1

The questions on the blackboard relate to the diagram at the top left in figure 1.3.[4] Mathematicians refer to this kind of diagram as a *graph*. Let's consider these four questions in detail. To begin, figure 1.4 is a reconstruction of the graph on the blackboard.

Question 1: Find the adjacency matrix A

Our graph has four vertices (the four little circles), inventively labeled 1, 2, 3, and 4. This means that the *adjacency matrix* A of our graph will be a 4×4 matrix: the number in the ith row and jth column of A will be equal to the

---

[4] O'Donnell's source for this set of problems was P. W. Kasteleyn, Graph theory and crystal physics, in *Graph Theory and Theoretical Physics*, ed. Frank Harary (Academic Press, New York, 1967), 63–65.

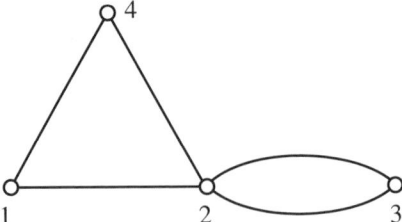

Fig. 1.4 The graph at the top of the blackboard in figure 1.3.

number of edges connecting vertex i with vertex j. So the adjacency matrix for our graph, which Will writes on the blackboard, is

$$A = \begin{pmatrix} 0 & 1 & 0 & 1 \\ 1 & 0 & 2 & 1 \\ 0 & 2 & 0 & 0 \\ 1 & 1 & 0 & 0 \end{pmatrix}.$$

For example, the two 2s in the matrix stand for the two edges connecting vertices 2 and 3.

Question 2: Find the matrix giving the number of 3-step walks

If n is any natural number, an n-step walk connecting two vertices is a journey from one of the vertices to the other that includes exactly n edges. (For this counting, an edge is counted as many times as you travel along it.) For example, the matrix A exactly records the numbers of 1-step walks between all the different pairs of vertices of the graph.

One elegant result in graph theory is that the number of n-step walks between the different vertices of the graph is similarly summarized by the nth power of A. Lambeau has instructed us to investigate the 3-step walks, and so the matrix we're after is

$$A^3 = A \times A \times A = \begin{pmatrix} 2 & 7 & 2 & 3 \\ 7 & 2 & 12 & 7 \\ 2 & 12 & 0 & 2 \\ 3 & 7 & 2 & 2 \end{pmatrix}.$$

Now, for example, the 2 in the top left corner of this matrix means that there are exactly two 3-step walks that start and end at vertex 1. These two walks consist of traveling clockwise and counterclockwise around the triangle.

Question 3: Find the generating function for walks from point $i \to j$

Let's consider two vertices, which we shall call i and j. Then for any natural number n we can count the number of n-walks from i to j; in the movie, Will denotes this number by $\omega_n(i \to j)$. For example, between the vertices 1 and 3 we can calculate that there are no 1-walks, two 2-walks, two 3-walks, and so on. The resulting sequence of numbers begins

$$0, 2, 2, 14, 18, 94, 146, 638, \ldots$$

The so-called *generating function* for this sequence is written as

$$\Gamma^\omega(p_1 \to p_3, z).$$

This stands for the infinite sum

$$\sum_{n=0}^{\infty} \omega_n(1 \to 3) z^n = 0 + 2z + 2z^2 + 14z^3 + 18z^4 + 94z^5 + 146z^6 + 638z^7 + \cdots.$$

From here on things get a bit more difficult, and we require some matrix determinants. Basically, what Question 3 is about is finding a way of calculating the generating functions that avoids calculating all those powers of the matrix A. If all this math is new or appears forbidding, there is no harm in your jumping ship and swimming to the next section.

The generating function $\Gamma^\omega(p_i \to p_j, z)$ is a function of z, and it turns out that it is always a *rational function* of z, that is, a ratio of two polynomials. This function turns out to be given by the formula

$$\sum_{n=0}^{\infty} \omega_n(i \to j) z^n = (-1)^{i+j} \frac{\det(\mathbf{1}_{ij} - zA_{ij})}{\det(\mathbf{1} - zA)}.$$

Here, the $\mathbf{1}$ stands for the 4×4 identity matrix, and $\mathbf{1}_{i,j} - zA_{i,j}$ is the (i, j)th minor of the matrix $\mathbf{1} - zA$.[5] All that seems to be required for Question 3 is to write down this formula. Note that Will's formula on the blackboard in figure 1.3 is missing the factor $(-1)^{i+j}$; as it happens, this does not affect Will's calculation, since he then applies the formula in the case $i + j = 4$, an even number.

---

[5] In case you're wondering where this formula comes from, here is a clue. Apply the formula $\sum_{n=0}^{\infty} z^n = \frac{1}{1-z}$ to the formal sum $\Gamma^\omega(p_i \to p_j, z) = \sum_{n=0}^{\infty} (A^n)_{ij} z^n = \left(\sum_{i=0}^{\infty} A^n z^n \right)_{i,j}$ and you get (don't worry about the details) $((1 - Az)^{-1})_{ij}$. Now, apply Cramer's rule to calculate the matrix inverse.

Question 4: Find the generating function for walks from point 1→3

So, Question 4 asks us to apply the formula in Question 3 to the special case of vertices 1 and 3. We calculate the generating function:

$$\frac{\begin{vmatrix} -z & 1 & -z \\ 0 & -2z & 0 \\ -z & -z & 1 \end{vmatrix}}{\begin{vmatrix} 1 & -z & 0 & -z \\ -z & 1 & -2z & -z \\ 0 & -2z & 1 & 0 \\ -z & -z & 0 & 1 \end{vmatrix}} = \frac{2z^2 + 2z^3}{1 - 7z^2 - 2z^3 + 4z^4}.$$

On the blackboard we can only see the denominator of the first fraction. But it seems Will gets it right, and his correct answer to the question is

$$2z^2 + 2z^3 + 14z^4 + 18z^5 + 94z^6 + \cdots.$$

1.8 Mathematics: Eigenvalues

Fig. 1.5 Lambeau in front of blackboards full of eigenvalues and eigenvectors.

O'DONNELL: "Basically, I followed the script in the sense that if they said they were going to talk about eigenvalues or something, I made up something with eigenvalues.[6] In the second big lecture hall scene I knew they were going

---

[6] Eigenvalues, often represented by the Greek letter λ, are special numbers associated with square matrices. They are visible on the blackboards in figure 1.5.

to start with him saying, after the mysterious person hadn't turned up, that there were a lot of people present that weren't in his class, and so if those people did not want three hours of eigenvalues and eigenvectors, they should leave.

"I took that just the way I'd teach a class here, to have nothing on the board to begin with. I would start the lecture with a blank board. But Gus said to me, 'We need something on the board.'

"So for that scene I had no notes made up at all, because I didn't expect I would be doing anything. I put a 3×3 matrix on the board that comes from dealing with three balls coupled with four strings, say one on either side. I wrote that down on the board and I wrote some garbage somewhere else, and on another board I used words like eigenvalues and eigenvectors all over the place. And then Gus came to me and said, 'No, we're taking a shot of him, and the board over his left shoulder is blank. We need that filled in, too.'

"So, if you look at that scene you see the words 'eigenvectors' and 'eigenvalues' everywhere because I just had to fill in the boards with something. But what we didn't expect was that the professor when he finished would come to the board and rub out the bottom board. That was not called for at that point.

"Usually the props manager takes a Polaroid of things in a scene. So that if they had to reshoot it, they know exactly what was there, a red coathanger beside a blue coathanger or whatever. But here he had done that so quickly in the first rehearsal that the props guy hadn't had a chance to take [the Polaroids] and I only remembered a bit of it, and of course I wasn't sure whether I could do the same thing again.

"But eventually it turned out that they had washed the boards before they had started, and so I could see a faint outline of what I had written, so I could fill in the matrix okay. And then I was about to leave when this girl came running out, one of the extras and she said, 'The last time you had $A =$ the matrix' [laughs]. We were saved because then we could take a photograph of it. I could do it each time."

1.9 Mathematics: Graph Theory 2

O'DONNELL: "I think in the script the director had made a reference that he was going to set a superhard problem. So, when I came across this paper by Harary on graph theory,[7] I realized that it was just perfect for people who can't write on the blackboard.

"These graphs, you could color them, use color chalk, for example. We decided the graph that Will could do best is the one that looks like the

---

[7] F. Harary and G. Prins, The number of homeomorphically irreducible trees, and other species, *Acta Mathematica* 101 (1959), 141–162. See, in particular, page 150 and Appendix II on page 161.

Fig. 1.6 The second set of 'superhard' problems has again been solved.

letters H and K stuck together,[8] and all he had to do was draw the line of
both the H and the K. I thought that covered both having a research type
problem and still overcoming this writing on the blackboard problem."

Let's have a close look at the two questions on the blackboard.

Question 1: How many trees are there with n labeled nodes?

Here, a *tree* means a connected graph with no loops. The question is then:
Jotting down n points on a piece of paper, how many essentially different
trees are there with these points as vertices? The answer, known as *Cayley's
theorem*, turns out to be n^{n-2}.[9] For example, figure 1.7 shows the $4^{4-2} = 16$
different types of trees that can span four points in the plane.

---

[8] This refers to the first diagram in the middle row of the blackboard in figure 1.6.

[9] The source for this question was J. W. Moon, Counting labeled trees, in *Lectures delivered to the Twelfth Biennial Seminar of the Canadian Mathematical Congress* (Van-
couver, 1969), Canadian Mathematical Monographs 1 (Canadian Mathematical Congress,
Montreal, 1970). See, in particular, the summary on page 3. The list of graphs on page 2
was also one of the items that O'Donnell had earmarked for use on the blackboard in this
scene.

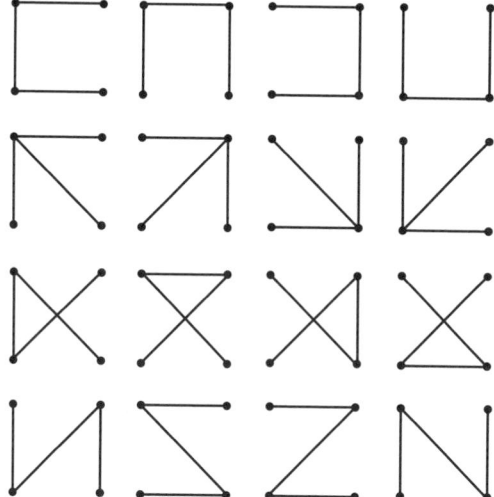

Fig. 1.7 The sixteen different spanning trees connecting four points in the plane.

Question 2: Draw all homeomorphically irreducible trees with $n = 10$ (vertices)

A homeomorphically irreducible tree is one in which each vertex has either one or more than three edges ending in it. Will is in the process of drawing an eighth tree when he is interrupted. The complete list consists of the ten trees pictured in figure 1.8.

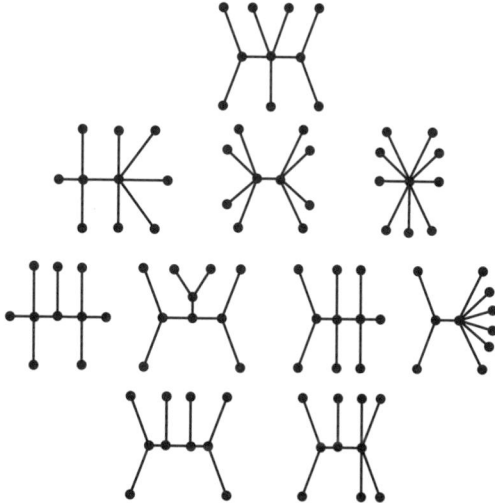

Fig. 1.8 The ten homeomorphically irreducible trees with ten vertices.

1.10 Mathematics: Graph Theory 3

0:26
Lambeau and Will are doing math together (figure 1.9).[10]

Fig. 1.9 Will calculates the chromatic polynomial of a graph.

Will completes the formula on the blackboard by adding the factor $(k-2)^6$ to the numerator, to arrive at

$$\frac{k^3(k \div 1)^3(k-2)^6}{k^2(k-1)^2(k-2)^2}.$$

Will and Lambeau take turns canceling common factors in the fraction, simplifying the expression.

What they are doing here is calculating the *chromatic polynomial* $P_G(k)$ of the graph G on the left of the blackboard. Here, $P_G(k)$ is defined as the number of ways to color the vertices of the graph with k colors, so that no two vertices in the graph connected by an edge have the same color. And, indeed,

$$P_G(k) = k(k-1)(k-2)^4$$

is correct, the result of their canceling extravaganza.

For example, this formula tells us that there are no legal colorings using only one color (well, duh) or only two colors. It also tells us that there

---

[10] This is an example taken from R. Brualdi, *Introductory Combinatorics* (third edition) (Prentice Hall, Englewood Cliffs, NJ, 1999), 514.

are six legal colorings using three colors. Figure 1.10 shows a legal color-
ing/numbering of the graph with three colors. The other possible three-
colorings are simply obtained by cycling through the six permutations of
the three colors.

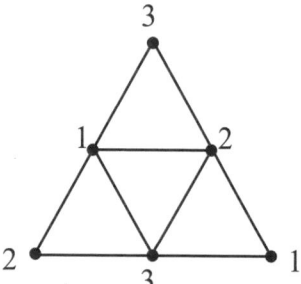

Fig. 1.10 Labeling with the numbers 1, 2, and 3 so that no edge has vertices with the
same number.

1.11 Mathematics: The Salieri Scene

O'DONNELL: "In the scene in which the young Will overthrows the famous
old 'Hungarian' [*figure 1.11*]—in the script they call him Alexander Pekec
[*perhaps a stand-in for the famous mathematician Paul Erdös?*]. I decided to
emulate Richard Feynman's 1948 meeting at the Oppenheimer conference,
beautifully described in Freeman Dyson's book *Disturbing the Universe*, in
which he describes Julian Schwinger and Dick Feynman as the young Turks
giving Neils Bohr and other worthies the new quantum electrodynamics.

"Although the immediate outcome of that meeting was the opposite of
that in the movie (Feynman was stunned and depressed by the response to
his way of doing QED without equations), the [*eventual*] outcome was indeed
the same—most physicists now use Feynman diagrams. I decided to make an
invention which looked like both graph theory/combinatorics and Feynman
diagrams, to counteract the 'old way' of heavy complicated mathematics. So,
you may see something vaguely familiar but with something wrong/irrelevant,
but have fun inventing this scene in your own way."[11]

Thus ends Patrick O'Donnell's wonderful telling of his excellent movie
adventure.

---

[11] The sources for the graphs and "complicated" equations that the simple "new" thing
is overthrowing were inspired by material from the book by J. W. Moon referenced in
footnote 9 on page 14: in particular, pages 58 and 59; and, from J. W. Moon, Graphical enumeration problems, in *Graph Theory and Theoretical Physics*, ed. Frank Harary
(Academic Press, London, 1967), 1–41.

Fig. 1.11. Tom (mathematician John Mighton) right,[12] with the upstaged professor.

1.12 Further References and Remarks

Saul, Mark, Movie Review of *Good Will Hunting*, *Notices of the American Mathematical Society* 45 (1998), 500–502.

Kleitman, Daniel J., My Career in the movies, Sidebar to Mark Saul's review of *Good Will Hunting*, *Notices of the American Mathematical Society* 45 (1998), 502. Daniel J. Kleitman is professor of mathematics at MIT. This short article tells of his role as math adviser for the initial script of the movie and his appearance in the background of one of the scenes. In fact, he walks by the window twice in this scene. First from right to left, then from left to right; see figure 1.12.

1.13 Parodies

Its story line makes *Good Will Hunting* ripe for parodying. A pretty good job is done in the episode "A Genius Among Us" (1999) of the Disney cartoon series *Recess*. A very quick but much funnier parody appears in the episode "Don's New Flatmate" of the British TV series *How Not to Live Your Life* (2009). Here, the doltish Eddie adds the (correct) single line $a = b$ to the tree-filled blackboard, which is later declared to be a great breakthrough (figure 1.13).

---

[12] John Mighton, who plays Lambeau's assistant Tom in the movie, was a graduate student in mathematics at the University of Toronto. Mighton is the mathematician behind the tremendously successful Canadian-based Junior Undiscovered Math Prodigies (JUMP) Project, which provides free tutoring to children with difficulty in mathematics. He is also an award-winning playwright.

Fig. 1.12. One last real mathematician, Daniel J. Kleitman, appears in the background.

Fig. 1.13. Eddie, about to do a Will Hunting.

Chapter 2
The Clever Hand Behind *A Beautiful Mind*

A Beautiful Mind (2001) tells the story of John Nash, mathematical genius and winner of the 1994 Nobel Prize for economics, and his struggle with schizophrenia. The movie is very unusual in that the main character is a living mathematician and in that the story is only loosely connected to actual events. Like *Good Will Hunting*, this is a movie more about psychology than mathematics. However, mathematics and mathematicians are portrayed in the movie in a way that rings true and amounts to a self-contained artistic piece within the larger whole. The man responsible for getting it so beautifully right is David Bayer, a professor of mathematics at Columbia University in New York City.

Our goal here, as in chapter 1 on *Good Will Hunting*, will be to provide something in the nature of a math consultant's commentary. It is based upon several sources: newspaper reports featuring David Bayer,[1] our own interview with him, and a recording of a conversation between Robert Osserman (then director of the Mathematical Sciences Research Institute), Sylvia Nassar,[2] and David Bayer.[3]

We'll stick to David Bayer's words as much as possible, adding screen shots, footnotes, and commentaries when they add some clarification. *As in the previous chapter, some of the commentary and discussion refers to high level mathematics. We encourage the reader to simply skip over any section that feels too heavy.*

---

[1] Blake Eskin, A beautiful hand, *The New Yorker*, March 11, 2002; Dana Mackenzie. Beautiful mind's math guru makes truth = beauty, *Science* 295 (2002), 789–791; Cynthia Wu, Math prof behind "A Beautiful Mind" speaks, *The Phoenix*, April 24, 2003; The ultimate backstage pass, *SIAM News*, December 13, 2001. At the time of writing, these articles were also available online.

[2] Author of the biography *A Beautiful Mind: A Biography of John Forbes Nash, Jr., Winner of the Nobel Prize in Economics 1994* (Simon & Schuster, New York, 1998).

[3] www.msri.org/publications/ln/msri/2002/non-workshop/nassar/1/

In 2000, David Bayer was asked to write a review of the Broadway play *Proof* for the *Notices of the American Mathematical Society*.[4] Ron Howard, who read and was impressed by this review, became the director of *A Beautiful Mind*. He invited David Bayer for an interview, which went so well that Howard hired him on the spot as the mathematics consultant for the movie. Bayer's role was multifarious: producing the writings on the blackboards and windows; contributing parts of the dialogue that involved mathematics; writing a program for the Pentagon scene, which was supposed to suggest how Nash's mind picks out patterns from seeming chaos; acting as Russell Crowe's hand double; being present as consultant whenever Crowe was acting in a math scene; consulting on the games of Go and Hex; and actually acting as one of the professors (the third in line) in the pen-presentation ceremony toward the end of the movie (figure 2.1). Between February and June 2001 Bayer spent hundreds of hours on this movie; given the very impressive results, many mathematicians, especially David Bayer himself, consider this time very well spent.

Fig. 2.1 Math consultant David Bayer (right), appearing in the pen-presentation scene.

Anyone interested in finding out more about John Nash and his mathematics should read Sylvia Nasar's excellent biography and the annotated collection of John Nash's (very difficult) papers, edited by Harold W. Kuhn and Sylvia Nasar.[5] The American Mathematical Society has also published

---

[4] David Bayer, Theater review of the play *Proof*, *Notices of the American Mathematical Society* 47 (2000), 1082–1084.

[5] Harold William Kuhn and Sylvia Nasar, eds., *The Essential John Nash* (Princeton University Press, Princeton, NJ, 2001).

some enlightening reviews of the movie and Nasar's book, from a mathematician's point of view.[6]

2.1 Real Math Beats Fake Math

David Bayer begins his story: "I was under the impression that I was given much more resources to inject math into the movie *A Beautiful Mind* [*in comparison to consultants working on other movies*]. Generally the point of view is that mathematics is inserted in specific scenes, whereas the motivation for Ron Howard coming to me for consulting was that you can't make up reality—that there will be a certain vérité to having as much math involved as possible, even if most people who watch the movie wouldn't understand a word of the math. I've had various people observe me doing math under various circumstances over the years and not understand a word of what I was doing, but still be drawing intense conclusions about my behavior.

"And in some sense, that's our view of what happens when you make the movie. You know, millions of people were in the theatre in a couple of hours' session of intense concentration and they are all observing, bringing different points of view to the observation. And you can't fake the reality that you would get from them actually giving me the time to do actual mathematics, even though it is on a board that is just going to be glanced at."

2.2 Signature Scene

BAYER: "Actually I brought some baggage with me, because I know of so many mathematicians who joke that they can solve the blackboards in *Good Will Hunting* while they are on screen. It is sort of fairly popular to discount it as easy graph theory, or whatever it is. So, there was a signature scene, which was analogous to that in *A Beautiful Mind*. Not a signature scene from the point where I contributed, but more when they came to me, this was forefront in their minds, something they were worried about. In the script, Russell Crowe gives some problem to this class which then Jennifer Connelly comes into his office to give a solution for, right?"

Setting a Trap

BAYER: "I was fearful when they asked me for this that people would be ripping whatever problem [*Bayer produced*], the same way they were ripping the *Good Will Hunting* scene. So I kind of rigged a trap with the problem. You know, basically, Ron Howard encouraged me to think like an actor, I mean he actually loves actors and everything he does is fundamentally an actor's

---

[6] L. M. Butler, Movie review of *A Beautiful Mind*, *Notices of the American Mathematical Society* 49 (2002), 455–457; J. Milnor, John Nash and *A Beautiful Mind*, *Notices of the American Mathematical Society* 45 (1998), 455–457.

Fig. 2.2 Nash sets a problem that will take months, or "the term of your natural lives."

medium. You can hear him discuss this point if you listen to his director's commentary on the DVD, sort of, in the first library scene where there is math. So, I put myself in the character the same way that I did everything else thereafter and thought, 'What is it about multivariable calculus that gets near something of deep interest?' And there is this one aspect of topology, which is the shapes of things.

"How can you measure when things are different shapes? There is a fact in multivariable calculus which is nearly true, but not quite true, and it fails if the space you are working on is a strange shape. In that exact era [*in which Nash worked*], DeRham was noticing it and first published papers that if you view this defect as an attribute, you could actually use it as a tool to measure the shapes of space. This is DeRham cohomology.

"So I thought about it, and the way this is scripted is that he walks into this room, and he really has not been thinking about teaching this class at all. He is just winging it, and what would I think about on the way over if I was winging a class like that? Why do you want to do this? What is interesting about this? This is the one fact that comes to mind. So the problem that he puts of the blackboard is effectively to calculate—I mean it's been so compressed to fit into the scene, it is almost cryptic. But basically it is a question about calculating the cohomology around some subset removed from $\mathbb{R}^3$.[7]

"So now, it turns out that the answer depends on what your category is. If you, say, have two lines which meet in a plane, and if you are thinking as an algebraic geometer, then you can't really have a pole around half of

---

[7] The problem Bayer is referring to can be seen in figure 2.2. The subset that gets removed from $\mathbb{R}^3$ is denoted by X.

a line without having a pole around the other half of the line. You get one answer, whereas if you think as an analyst, you *can* certainly have partitions of unity, and have something really wacky happen on one half and not have it happen on the other half, and you get a different answer. And actually one of the algebraic geometers at Columbia gave as I was expecting the algebraic geometry answer. So basically I was trying to trap anybody, and I think that many people actually took the bait."

Acting Mathematics

BAYER: "Russell had actually been very nervous the day that he did the math lecture, where he wrote the problem out and talked to Jennifer at the same time in the classroom. And everybody was sitting there astounded: 'You mathematicians really talk and write at the same time?' Here we are taking like twenty takes and no one was thinking that Russell was a yahoo and everyone is extremely impressed with it. They were basically feeling sympathetic with him that he was trying to pull that scene and they are looking at me: 'You guys do this? You gotta be kidding.'

"Oh, by the way, they faked the book that is thrown into the wastebasket with my name as the author on it [Calculus of Several Variables *by D.A. Bayer*].[8] This was a standard trick because you want to make sure that someone won't sue. Russell realized halfway through that scene that he is throwing my book into the trash. He acted momentarily like he was putting in a little bit too much zeal. He had a bet running with the prop guy that he could make the shot without hitting the basketball rim. He realizes that he would trash my book and he gave me this funny quizzical look: 'Are you okay with this?' 'Yeah it's a cameo, it's okay. Go.' "

On the Importance of Getting It Right

BAYER: "I hung around whenever there was math involved. It was almost like it was an entitlement that whenever Russell Crowe was on the set and there was possibly math in the scene I would be available, in case he wanted to ask me a question. Whenever the prop department wanted to decorate the scene with stray math they wanted me around to make stuff fresh for them. So that made me be around an awful lot.

"Russell Crowe asked me to come in during the scene in the office with Jennifer Connelly [*in which Connelly shows her solution to the problem*], to point to where on the page he would be glancing at to know that she had done it wrong, and I basically pointed at the middle of the page. Then Jennifer Connelly looks at me and asks me pretty much in these words, 'Are you making me look like a yahoo here? How bad is my answer?' And I explained,

---

[8] Another scene in which David Bayer's name is playfully mentioned is when John says 'Papers in hand, Mr. Bayer' to one of the students exiting his classroom.

'Actually it's a pretty high-level mistake. I got a Columbia professor to make the same mistake.' She liked that."

Blackboards and Windows

BAYER: "Certainly one of the more stressful points for me was just early on. The script has the scene where we are doodling on the library window. There are three panes, one of pigeons in the park and the third one is of observing a mugging. And I had to decide how to decorate the window [*figure 2.3*].

Fig. 2.3 Window doodles, of a pigeon cluster (left) and a purse being stolen (right).

"There is no really literal answer here, in that it was a bit of a stretch even in character to imagine John Nash doing it exactly as it was portrayed dramatically here. And what would the math look like of analyzing pigeons in the park, which was actually a little bit different from game theory, but it might be game theory. I struggled with it, and Ron wanted to see what I was going to do, and it would ruin a day's filming if I didn't do something. He was getting nervous as it was getting close: 'Hey Dave go use my trailer.

I want to see something in four hours. Nothing is going to happen here, you know.'

"And ultimately I ended up actually writing out a storyline for myself what was taking place. I started translating it into graphs and then I started putting values like I was writing a game theory puzzle on the graph. And even though I don't know of game theorists doing exactly this to such a storyline, that was the basis for my drawing. My drawing, if you look at it close enough is actually quite literal [*see the right closeup in figure 2.3*]. It is some story of a mugging and that's what they wanted. I think there is a tendency for mathematicians to be somewhat dogmatic and too literal minded. There is a kind of relaxation that had to take place for me to pull that off."

2.3 The Riemann Hypothesis

BAYER: "In the second part of the movie, rather than being realistic to what happened to John Nash and his math, which was scattered, they wanted to have a consistent storyline where he'd make incremental progress toward the Riemann hypothesis."[9]

Riemann on the Porch

BAYER: "The first moment where you see this is when he is sitting on the porch and Saul comes by, he doesn't want to take his pills in public and he's got this clipboard. He hands Saul the clipboard (figure 2.4), Saul glances at it and says: 'Well, you know, maybe there are other things in life besides math.' You actually do a closeup on the clipboard a little bit long and John Nash wrote me an email: 'What was on that clipboard?'

"It was supposed to be in some sense numerological gibberish, but he had the sense that it wasn't quite gibberish, and he wanted me to tell him what it was. And he was right, it wasn't quite gibberish. I tried to think wildly, but in character. And something that fascinates some people, and may not be a mainstream approach to the Riemann hypothesis, sort of struck me as relevant. There is the way continued fractions behave differently than our usual view of numbers. Well actually there was a bunch of numerology involved in continued fractions on that sheet of paper."[10]

---

[9] The *Riemann hypothesis* is one of the most important unsolved problems in mathematics. It is concerned with the zeros of the so-called *Riemann zeta function*. It is one of the Clay Mathematics Institute Millennium Prize problems. Proving that the Riemann hypothesis is true would give major insights into the distribution of prime numbers. See figures 2.5 and 2.6 for more details about the zeta function, and the connection between the Riemann hypothesis and prime numbers.

[10] The expression $1.4142135623\ldots$ is the start of the well known infinite decimal expansion of $\sqrt{2}$. Another way of writing $\sqrt{2}$, familiar to mathematicians if not more generally, is the infinite *continued fraction expansion* (see next page)

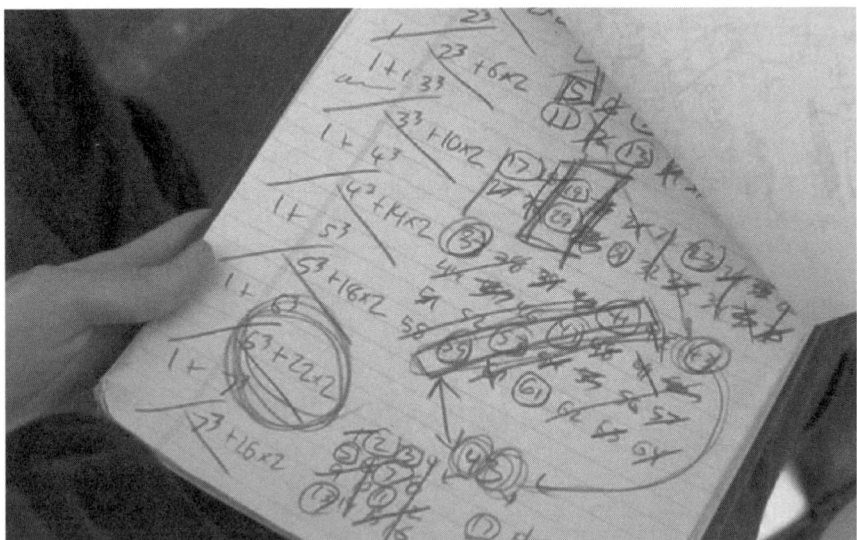

Fig. 2.4 There are two pages that Saul looks at on the porch. The first (not shown) is filled with what appear to be random numbers, and what John is obviously trying to do is pick out some patterns. The writing on the second page (shown above) appears more coherent: the expression on the left looks like a somewhat unusual way of writing a continued fraction. Part of what is happening on the right looks a little bit like the first few prime numbers (circled) being generated using *the Sieve of Eratosthenes*, or attempts to discover something about the prime numbers, by highlighting patterns among the primes and among the nonprime integers.   .

Riemann in the Lecture Theatre

BAYER: "Then we transport the Columbia lecture to Harvard, for the scene where he breaks down and gets taken to the asylum. I realized that the three boards I was filling there were likely to be scrutinized closely by number theorists, and therefore I tried to put a stutter step in the first board [*figure 2.5*], so that anybody who was reading it really closely would just be thrown on board one and would just run out of time before they got to board two [*figure 2.6*] or board three. At least until they get the DVD.

"He starts saying something which the audience thinks is crazy, something along the lines of the zeros of the Riemann zeta function corresponding to singularities in space-time. Okay, this is a very polarizing comment. First of all the quaternions generalize the complex numbers, and who knows maybe

$$1 + \cfrac{1}{2 + \cfrac{1}{2 + \cfrac{1}{2 + \cfrac{1}{2 + \cdots}}}}.$$

All real numbers can be expressed as both decimal numbers and continued fractions. The continued fraction expansion of a number is actually the more natural, usually telling mathematicians a lot more about the number.

Fig. 2.5 A closeup of the first of the three blackboards in the Columbia lecture. At the very top are the two most famous ways of expressing the Riemann zeta function $\zeta(s)$. The quantity $s = \sigma + it$ is a complex variable, indicated at the top right. The zeta function is zero when $s = -2, -4, -6, \ldots$. The *Riemann hypothesis* asserts that all other zeros have real part $\sigma = \frac{1}{2}$. The infinite sum $\sum_{n=1}^{\infty} \frac{1}{n^s}$ at the top left also makes a later appearance; there, John is riding his bicycle in the form of an infinity sign which transforms into the infinity of the sum.

One connection of the Riemann zeta function with prime numbers is visible in the expression $\prod_p (1 - \frac{1}{p^s})^{-1}$, which is known as the *Euler product formula*. This is an infinite product in which p stands for the prime numbers. The second line is the equation $\Gamma\left(\frac{s}{2}\right) \pi^{-\frac{s}{2}} \zeta(s) = \Gamma\left(\frac{1-s}{2}\right) \pi^{-\frac{1-s}{2}} \zeta(1 - s)$, which captures an important symmetry of the zeta function. The third line involves another very prime-heavy function, the so-called *Möbius function* μ. The stutter step mentioned by David Bayer, involving a weird continued fraction expansion, is visible at the bottom of the blackboard.

it would be profitable to reconsider the Riemann hypothesis for the quaternions. It might be easier if you generalize it properly.[11] In any case, theoretical physics already got into the character's mind and there is this loose connection.

---

[11] The quaternions are numbers that extend the complex numbers similarly to the way in which the complex numbers extend the real numbers. The idea that Bayer is putting into Nash's mind is in a sense very natural: a problem that is difficult when expressed in terms of complex numbers may turn out to be much simpler when thought of in terms of quaternions. This process of "simplifying by generalizing" is very common in mathematics. A famous example is the solution of polynomial equations in terms of complex rather than real numbers.

"I view insanity as a partial gift, in that it allows one to make looser formal associations, that you are making an analogy between space-time and quaternions was the basis for making up that line. I tried it on Brian Green, who I greatly respect, he is a friend of mine at Columbia. He gave me the wildest look when I tried that line on him, and the minute I saw that look I thought 'Okay, that has to be the line in the movie.' But of course various people trashed that. It's not correct. It is not correct mathematics, but it is in character."

Fig. 2.6 A closeup of the second of the three blackboards in the Columbia lecture. This one is concerned with the connection between the Riemann hypothesis and the distribution of the prime numbers. The important bits, in the upper left corner, are $\pi(x)$ (the number of primes less than or equal to x), $\mathrm{li}(x)$ ($= \int_2^x \frac{dt}{\ln t}$), and the expression $O(x^{1/2} \log x)$. The Riemann hypothesis being true is equivalent to the second equation being true. This second equation says that the complicated function $\pi(x)$ is approximated by the simpler function $\mathrm{li}(x)$, within an accuracy of $O(x^{1/2} \log x)$. That is, the Riemann hypothesis being true is equivalent to there existing a constant C such that $|\pi(x) - \mathrm{li}(x)| \leqslant C x^{1/2} \log x$ for all positive numbers x.

Riemann in the Library

BAYER: "There is finally the board which probably had the most scrutiny in this theme, when he is in the library and the young kid comes and sees him and figures he is working on the Riemann hypothesis [*figure 2.7*]. There I needed to do something that could possibly be stared at in great length.

"I used two devices. First of all, Deligne had made progress over finite fields during that exact decade, and there are even to this day some number theorists who think that is a not a bad parallel for how you would approach the general problem, that one should work with noncommutative rings, etc.,[12] and basically what I did was make a very free-association transcription of Deligne's work.

Fig. 2.7 A student talks to John in the library.

"There are functions on the board whose definitions don't appear, so that if the blackboard was computer code you couldn't compile it, because you wouldn't know certain definitions. Which means, in particular, you could never prove me wrong. What it is intended to be for that moment is plausible progress for the proof of the Riemann hypothesis, which I found challenging to come up with (and I got help from various people). It was funny, if I knew of a proof of the Riemann hypothesis there is a thing called the Clay Prize, so Akiva Goldman the writer kept joking with me about that theme: 'So Dave, if you figure the Riemann hypothesis out, this is where you can put it in the film' [*laughs*].

"The part that is personal about that for me is that when I was a graduate student, at first-year graduate school at Harvard. I was taking Galois theory with Barry Mazur, and Graeme Segal was visiting. And he taught topology, and it kind of blew my mind when I first realized that somehow covering

---

[12] This is a similar idea to that mentioned in the previous footnote, regarding the quaternion numbers. The new "numbers" here are the so-called *finite fields*. If you succeed in finding the solution in the new setting of finite fields, you can then hope to solve, or at least gain insight into, the original problem.

spaces for topological spaces and Galois extension fields were fundamentally the same thing.

"This is sort of this moment as a graduate student, you know the way you are when you are a first year graduate student, and things are just amazing to you. This is one of the things that was amazing to me, and so I put those words in the student's mouth who approaches Nash. Except that in the scene, the student feels that he has thought of this by himself and does not realize that we already know this.

"In a way if someone is coming uncharitably at the film, then it is a challenging delivery because the actor did a great job of being right on the edge of like, you know, 'Is this guy a prodigy or wacko or what the hell?' And his delivery, you are not sure how you are supposed to read it if you trying to sit there and break down the math.

"But my point of view was that I don't care. And basically I'll do things in character, and if it is in character to risk drawing criticism, I knew that the character drawing criticism was happy to take it. I see myself as having done my job if I can have a character say something like that."

Riemann Erased

BAYER: "I had to come up with some kind of calculation related to the Riemann hypothesis, but you know something I could do naturally myself. It was way too involved for me to do from plans. I basically just ensconced myself in the room for six hours and started scribbling on boards, trying calculations and looking for early visual calculational tools—I went around the room about six times, and every time I went back I would use spaces in between what I had done before, erase something more, and so built up this pastiche.

"So, Ron had to go into the room for a different reason, looking through the window down into a scene they were shooting. He'd come back, he's thrilled with this. We get into the room in the afternoon finally and Russell looks around, and according to the script he is supposed to go and add a line or two to it. However, this did not feel right to Russell. 'I think what I should do is look at this from the hallway, do a lot of reflection, then turn around and erase the board' and he grins: 'I would have wrecked Dave Bayer's whole afternoon' [*figure 2.8*].

"So Ron, basically now he is thinking cinematographer, because in any given take the guy holding the focus ring can get it wrong, or there can be dots on the negative, or whatever. So, basically, 'I tell you what, we do a couple of takes where you just go up to the board—and then we'll do a take where you erase.'

"Now I had to get out of the room, it's a tiny room. They do a few practice runs and they are like shouting back and forth: 'Can you erase?' 'Yeah.' 'Can you erase?' 'Yeah.' You know that's funny. Finally they go for it and then Ron comes bounding up the stairs and he says, 'Dave how long would it take

Fig. 2.8 Six hours of David Bayer's work about to be erased.

to do that again?' It was a really visibly pulled punch. He did not want me for an instant to think he was serious, that he didn't appreciate that it actually did take me six hours. But it was so funny everyone cracked up."

2.4 The Blonde and the Nobel Prize

A pivotal moment of the movie is the bar scene, in which a blonde bombshell and her friends enter the bar, and John has a brainstorm:

0:19
JOHN: Adam Smith needs revision.
MARTIN: What are you talking about?
JOHN: If we all go for the blonde, we block each other. Not a single one of us is going to get her. So, then we go for her friends, but they will all give us the cold shoulder because nobody likes to be second choice. Well, what if no one goes for the blonde? We don't get in each other's way, and we don't insult the other girls. That's the only way we win. That's the only way we all get laid. Adam Smith said: "The best result comes from everyone in the group doing what's best for himself." Right, that's what he said. Incomplete. Okay? Because the best result will come from everyone in the group doing what's best for himself and the group.
MARTIN: If this is some way for you to get the blonde, go to hell.
JOHN: Governing dynamics. Adam Smith was wrong.

At this point Nash rushes off. What he had supposedly hit upon are the fundamentally new ideas about noncooperative games that were the subject of the (real) Nash's PhD thesis. Mathematicians use noncooperative games to model real-life scenarios such as arms races, the stock market, and so on. Nash's ideas proved extremely influential, enough to earn him the Nobel Prize for economics in 1994.

To get a sense of the mathematics intended to be behind this scene, let's first have a look at a simple example of a noncooperative game.[13] In this game, three players try to win money from a pot. For each play of the game, the players simultaneously stick out 1 or 2 fingers. Then the rules are:

- If only one player sticks out 1 finger, that player wins $1 from the pot.
- If only one player sticks out 2 fingers, that player wins $4 from the pot.
- In all other cases, nobody wins or loses anything.

The four possible finger combinations (up to rearranging the players) are

$$(1, 1, 1), (2, 2, 2), (1, 2, 2), (2, 1, 1).$$

In his PhD thesis, Nash introduced and investigated the concept of what is now called a *Nash equilibrium* of a noncooperative game. This is an outcome of a game that leaves all players "happy," or at least "not unhappy," in the following sense: none of the players could have increased their winnings by *unilaterally* changing their decision, with the other players leaving their decisions unchanged.

In our finger game, (2,1,1) is a Nash equilibrium: certainly, the person playing 2 fingers could not have done better; also, if either of the players who played 1 finger switched to 2 fingers, she would still not get anything, and so would not do better. Similarly, (1,2,2) is a Nash equilibrium. On the other hand, (1,1,1) and (2,2,2) are clearly not Nash equilibria.

It turns out that not every noncooperative game has such a simple equilibrium. However, things change for the better (mathematically) if we play such a game repeatedly and consider different *mixed strategies*. A mixed strategy is one where each player chooses from their options randomly, but with certain fixed probabilities. By including mixed strategies, we can then consider Nash equilibria for repeated games, and Nash proved that such equilibria *always* exist. In our game, one such equilibrium corresponds to all three players going for 1 finger $\frac{1}{3}$ of the time, and going for 2 fingers the other $\frac{2}{3}$ of the time.[14]

---

[13] The following is a variation of a game that Robert Osserman mentions in his conversation with Sylvia Nassar and David Bayer; see footnote 3.

[14] The probability $p = \frac{1}{3}$ is the unique solution between 0 and 1 of the equation

$$1 \cdot (1 - p)^2 = 4 \cdot p^2 \,.$$

Then, if each of the players plays 1 finger with probability p and 2 fingers with probability $1 - p$, the expected winnings of each player are $\$\frac{4}{9} \approx \$0.44$ per play of the game:

$$\$1 \cdot (1 - p)^2 p + \$4 \cdot p^2 (1 - p) = \$0.44 \,.$$

If then just one of the players changes his probability from p to q, it is easy to see that this player's average winnings will remain the same:

Now, how does all this fit in with what is going on in the scene with the blonde? Here are David Bayer's thoughts.

BAYER: "The bar scene is not considered to be a very good example of game theory, nor is it considered a good example of specifically explaining the Nash equilibrium. Harold Kuhn, who is a contemporary of Nash and a good friend of his, and also served as an advisor [*on the movie*] had been going back and forth, trying to convince Akiva to rewrite that scene completely.

"Harold Kuhn's explanation, while mathematically far more sound, served to illustrate why he is a mathematician and not a screenwriter. It was taking all that Akiva had put into the scene out of the scene. When I came on board there was this kind of sheepishness: 'Okay, when you look at the script there is this bar scene. Now, we know it's wrong. Can you stay on the job if this is the scene?' I didn't choose to fight that fight. I actually thought about it and I liked the scene in the movie. For me, everything I was doing was in character.

"John Milnor.[15] He came five minutes late to his first class for graph theory as a freshman, and there was a problem on the board. So he dutifully came into the next class and I think, as the legend goes, he said 'I think I am not cut out for this after all.' Because [*the problem*] was a lot harder than he thought it was, but he got it. Of course, it was an open problem and it became his first paper.

"This is kind of in the legendary landscape in Nash's mind, in that Nash in the movie came gradually to his final understanding of Nash equilibrium. But Nash, like anybody, would love to have this story be told about him.

"You can say that in some sense Akiva is channeling these legends. It is a dream, a fantasy. In a dream sequence Akiva is definitely channeling something on it for real, in having Nash have this insight in the bar. Although we very carefully back it up by showing his laborious pain working in his room, as winter turns to spring, to make it clear that he didn't just do it in one night."

As David Bayer says, there seems to be only a very loose connection between what is happening in the bar scene and the mathematical ideas that it supposedly inspires. Certainly, fighting for the blonde will likely be a fairly noncooperative game, and trying to figure out what the optimal strategy is seems to be the right approach in this context. However, a Nash equilibrium is not necessarily equivalent to an "optimal strategy" in the sense that John uses it in this scene. Furthermore, it seems impossible to interpret this scene as a noncooperative game having John's solutions as Nash equilibria.[16] Bayer

---

$$\$1 \cdot (1-p)^2 q + \$4 \cdot p^2(1-q) = \$0.44\,.$$

This confirms that we are actually dealing with a Nash equilibrium.

[15] In the movie, John's friend and competitor is Martin Hansen, who is modeled on the famous mathematician John Milnor.

[16] One attempt at a mathematical interpretation can be found in Lynne M. Butler's review of *A Beautiful Mind* in the *Notices of the American Mathematical Society*; see footnote 6.

is not so bothered by this, but our feelings are mixed: the bar scene is cinematically a great scene, but we find the mathematical inaccuracy distracting.

Back to the Pentagon Story

BAYER: "In any case pretty soon after this we are shooting the Pentagon scene, and there are these latitudes and longitudes along the US-Canadian border. Now, Akiva was basically writing nonstop, writing in the middle of the night, rewriting even though the script was nominally finished before they started. And Russell always had new ideas, and Ron always had new ideas. And the usual tension, we don't want the writer on the set because the director wants to maintain control, goes out the window.

"Akiva and Ron had a great working relationship, particularly with Russell. And what it meant was that Akiva would be on the set helping the director and then, instead of sleeping he was doing rewrites [*so that*] by the end of the movie it looked like he's had no sleep for three months. But I was asking him: 'Look, I have been helping Art [*the art department*]. We are gonna eventually need to know these names of towns on the US-Canadian border.' And he is looking at me like tired: 'Dave, don't worry about it.'

"And I gradually get drawn into it, where basically I was sick of this entire scene, where he said 'Dave could you look for towns on the US-Canadian border?' I had some GPS software on my computer and I could basically scroll along the entire border and pick out interesting names, like Starky Corner. And they ended up in the script, complete with coordinates.

"You may not think this is math, but it is math in a way. It is numbers, math is numbers in a movie. If *Drowning by Numbers* is math this is math, right? Basically Russell got used to the idea that a lot of movies just make stuff up. And at some point, Akiva is still building trust with Russell, he is like looking at Akiva like, 'There are some funny names in this script: does Starky Corner's name really exist, or is this bullshit?' And so Akiva comes to me like, 'Dave, Starky Corner.' He tells me what Russell said. 'Okay, Akiva I've got one of these math books on Maine in the car. I can bring it out and show you.' 'Yeah, please go get it.' So basically we are proving to Russell that this place actually exists."

2.5 Hand Double

BAYER: "The hand double bit came about by accident. Russell, Ron, and Akiva were discussing various scenes in rehearsal, and they had scheduled me to come in to talk to them about the stuff that was going to go on the blackboard. Russell would be writing it and I would be writing it and they looked at each other and said—I guess this was far more planned than I had any idea—that they realized that someone is going to do the math. And wondering whether I could do it, and they just wanted to approach it a certain way.

"Our handwriting looked pretty similar and our hands look pretty similar. I ended having to wear false nails for three months, I hated that. The executive producer said if anything demonstrates my dedication to this project it is that. I just clipped my nails back too far and he wanted to have longer nails.

"So basically if you are the hand double you are the hand double for everything. Where I would go to wardrobe, I would go to makeup, throw the gun into the river, moving markers on the board. It was kind of a Tom Cruise fantasy to be on the set, although they were struck that I tended to do everything in two takes and that was it. They imagined that I would be more nervous. They don't realize how us mathematicians have strange ideas like, 'We should have been Tom Cruise but for some genetic quirk,' and this is just our fate. It's not a big deal, it's nothing to be nervous about. Anyway that whole part was a lot of fun."

2.6 Bits and Pieces

$\alpha\beta - \beta\alpha \to \heartsuit$

Question: They get married and they are driving away in this car and on the rear window you can see the formula $\alpha\beta - \beta\alpha \to \heartsuit$. What was that all about?

BAYER: "Oh, I'm glad you are asking, I love that one. The script supervisor Eva Cabrera, it was not really her role to be asking this. But on the other hand, [*given*] the kind of positive environment, she thought that there should be something drawn on the back of the window of the car in the wedding scene, and no one had thought about it.

"She mentioned it to Akiva and he said, 'Yeah, go ahead, ask Dave.' He thought we should obviously be writing some math on the window and so the math I came up with was to have a commutator approach zero,[17] the idea that people would be more compatible. To mathematicians that would suggest increasing compatibility as this commutator would approach zero.

"I went through the whole script with Akiva and Ron in preproduction. And they wanted me to tell them everything that came to mind, even though they would take up only one suggestion out of ten. There are a lot of little things where Akiva and I were trying to write things together. And he would hear it, and both wanted me to believe the math sounded right, but also [*they kept in mind*] how it would sound to a nonmathematical audience. Because sometimes it could have the opposite sound.

---

[17] A commutator of two "numbers" α and β is simply the expression $\alpha\beta - \beta\alpha$. If both numbers are real numbers, then this commutator is always equal to 0, because real numbers "commute." However, there are other types of numbers in mathematics (for example, the quaternions that we mentioned earlier) that may not commute. So the commutator of such numbers could be different from 0. The commutator then "measures" how far away the two numbers are from commuting.

"Ron Howard and Akiva Goldman are very deliberate artists, very self-aware of what they are doing, and the fact is that they saw coming anything any critic could ever say. Making the movie was a deliberate artistic act, it brings in the money, it works, it makes people happy. So, in any way this is a perfect example of this, because zero has the wrong connotation. Akiva said, 'It can't be zero, but can there be a heart?' 'Yeah!' So it's very simple, we replace zero by a heart and everyone is happy."

$0 \leqslant \pi \leqslant 1$

BAYER: "This is again about you never get to choose the one percent. Many, many production stills, in fact whenever you now see an image from the movie, almost always, even years later, involve some math that I did. One of the ones that got a lot of flack, at the time [*was the one where*] Russell was looking through his dorm room window, and he is writing $0 \leqslant \pi \leqslant 1$.

"Again it is totally in character. I was immersing myself in John Nash's work and there is a paper he wrote on poker as a graduate student, in which he rather gleefully made sure that he needed exactly 24 symbols, and used all the Greek letters.[18]

"If you can say that each letter has a certain lock on a certain meaning, my elitist mathematical view [*laughs*] is that it is only swimmers at the shallow end of the pool who would ever draw the conclusion that π has anything to do with the circle. The rest of us overload every symbol we can, and this is what Nash was doing. He was overloading the symbols: $0 \leqslant \pi \leqslant 1$. I was basically using papers from that era to decorate the dorm room, to try to be as realistic as possible, but in the end people jump to conclusions."

Fly Versus Train

1:53
John discusses a very famous mathematical puzzle with his students.

JOHN: ... coming together at a maximum speed of, let us say, 10 miles per hour. So you have a fly on the tire of bicycle B, and the fly, who can travel at 20 miles an hour, leaves the tire of B and flies to the tire of bicycle A and backward and forward, until the two bikes collide and the poor little fly is squashed.

This is the important thing about actually focusing in, and comprehending the area you're dealing with. Mathematics is very specific, and it is an art form, no matter what people here will tell you, especially people from biology. Don't listen to those people. Let me go back to what you were doing, I might want to steal this, write a book and get famous.

---

[18] In fact, John Nash discusses this in a conversation with Ron Howard included on the DVD.

The information missing from John's description is how far the bicycles begin apart—let's just make up a number, say 1 mile—and the actual question to be answered: how far the fly will have traveled before it gets squashed?

Let's solve this problem, assuming both bicycles are traveling at 5 miles per hour. Since the fly is traveling at 20 miles per hour, its speed relative to bicycle A is 25 miles per hour. So, leaving bicycle B, the fly will reach bicycle A in $\frac{1}{25}$ of an hour, having traveled $\frac{20}{25} = \frac{4}{5}$ of a mile. In the same period, each bicycle will have traveled $\frac{1}{5}$ of a mile, leaving a distance of $\frac{3}{5}$ of a mile between them.

Having reached bicycle A, the fly then turns around, and heads back to B. To do this, the fly travels a further $\frac{4}{5} \times \frac{3}{5} = \frac{12}{5}$ of a mile, with the bicycles ending up $\left(\frac{3}{5}\right)^2$ of a mile apart. Continuing in this way, the total distance traveled by the fly is given by the infinite geometric sum

$$\frac{4}{5} + \left[\frac{4}{5} \times \frac{3}{5}\right] + \left[\frac{4}{5} \times \left(\frac{3}{5}\right)^2\right] + \cdots.$$

By a standard trick, this sum can be calculated to be 2. So, we conclude that the fly travels two miles before it gets squashed.[19]

There is actually a much simpler way to solve the problem. We only have to realize that it will take the bicycles one tenth of an hour before they collide. Therefore, the fly will have traveled $20 \times \frac{1}{10} = 2$ miles before this happens.[20] To misdirect people from this simple solution, rather than specifying that the bicycles are coming together at a certain speed (here, 10 miles per hour), the problem is usually posed by stating that A and B are traveling at specific speeds, say 3 miles per hour and 7 miles per hour.

Here are two very famous stories about the brilliant mathematician John von Neumann. Von Neumann was one of the inventors of game theory, and it was one of his theorems that Nash generalized in his PhD thesis. Nash's advisor in the movie is modeled on von Neumann.[21]

The first story goes that, when Nash shows von Neumann his proof of the existence of Nash equilibria, von Neumann is not at all impressed in the manner of the professor in the movie. Von Neumann supposedly more or less trivialized Nash's result, declaring it to be a simple fixed point theorem.

The second story is about a mathematical colleague posing the fly problem to von Neumann. Von Neumann comes up with the correct answer almost instantaneously. 'Interesting,' the impressed colleague is supposed to have remarked. 'Most people try to do the infinite sum.' 'What do you mean?' von Neumann replied. 'That's what I did.'

---

[19] If the speeds of the bicycles are not the same, the same ideas still work. However, the calculations are messier, involving two geometric sums.

[20] This method also makes it obvious that the final answer does not depend upon the individual speeds of the bicycles, only the sum of their speeds.

[21] John von Neumann also appears briefly as a character in *Race for the Bomb* (1987).

Chapter 3
Escalante Stands and Delivers

This is one of our favorites. *Stand and Deliver* (1988) is the story of Jaime Escalante, a mathematics teacher from Bolivia who comes to Garfield High, a poor and poorly run school in the Latino area of East Los Angeles. He is a humorous, charismatic, and demanding teacher, who persuades his class to take the Advanced Placement exam in mathematics.[1] The class performs so well that they are falsely accused of cheating by the testing authorities.

The story is substantially true, with important qualifications that we detail at the end of this chapter.[2] The real-life Escalante was made famous by the events at Garfield High, which culminated in his winning the Presidential Medal for Excellence in Education in 1988 and being inducted into the National Teachers' Hall of Fame in 1999.

Sadly, Jaime Escalante died in 2010. Right up until his death, Escalante was actively involved in promoting mathematics, in sharing his passion for mathematics. *Stand and Deliver* succeeds primarily because Edward James Olmos, who plays Escalante, captures this passion.

Stand and Deliver is not a story about mathematics, nor even really a story about mathematics teaching; it is a story of the human spirit.[3] Nonetheless, the movie contains lots of mathematics. Moreover, unlike the vast majority of school-based movies, the mathematics is not only internally correct, it is contextually correct: the mathematics the students fight with is what they would have had to fight with.

As the students progress, the mathematics progresses, from arithmetic and algebra to trigonometry and on to calculus. We won't comment on every

---

[1] The Advanced Placement Program is a scheme by which students in the United States can gain university credit for study undertaken in high school. The AP subjects are usually taught at school, but the exams are administered and independently assessed by the College Board, who are effectively the villains in *Stand and Deliver*.

[2] For a clear and thoughtful telling of the story, see Jay Matthews, *Escalante: The Best Teacher in the World* (Henry Holt, New York, 1988).

[3] Rare among human spirit movies, *Stand and Deliver* doesn't make us embarrassed to write "human spirit."

piece of math that occurs, but certain scenes are notable, giving the flavor of Escalante's style and the role it plays in the drama. These scenes will be our guide to the movie.

3.1 Welcome to the Finger Man

When Escalante first appears, he is confronted by a wild and totally disinterested class: *I don't need no math. I got a solar calculator with my dozen donuts.* The class is then ended by a premature bell, but the next class sets the tone. Escalante appears wearing a butcher's apron and wielding a huge cleaver. Chopping apples, he illustrates fractions to the students. Two cholos, Chuco and Angel, appear late, and Escalante enters into a whispered discussion with Chuco:

0:10
ESCALANTE: You know the times tables?
CHUCO [sticking up his thumb]: *I know the ones* [sticking up his second finger] *the twos* [sticking up his middle finger, in an obscene gesture], *the threes . . .*
ESCALANTE: Finger Man. I heard about you. Are you the Finger Man? I'm the Finger Man too. You know what I can do? [He holds his ten fingers splayed.] *I know how to multiply by 9. 9 times 3: 1, 2, 3* [counting off with his fingers, to leave two fingers on one side of his crooked finger, and seven on the other side; see figure 3.1]. *Whaddyou got?* [wiggling his fingers] *27! 6 times 9: 1, 2, 3, 4, 5, 6. Whaddyou got?* [wiggling his fingers] *54! Yeah. Want a hard one? How about 8 times 9? 1, 2, 3, 4, 5, 6, 7, 8. Whaddyou got?* [wiggling his fingers] *72!*

Escalante has demonstrated a clever trick for multiplying by 9 (which works because the sum of the digits in the answer is equal to 9); the trick is cute in itself, but it also contains a message relevant to the drama to come.[4]

3.2 Filling the Hole

In his next class Escalante explains zero, and negative numbers.

0:15
ESCALANTE: You ever dig a hole? The sand that comes out of the hole, that's a positive. The hole is a negative. That's it. Simple. Anybody can do it. Minus two plus two—equals—[He cajoles Angel into answering]. *You going to let these burros laugh at you?*
ANGEL [sullenly]: *Zero.*
ESCALANTE: Zero. You're right. Simple. That's it! Minus two plus two equals zero. He just filled the hole. Did you know that neither the Greeks nor

---

[4] The *Columbo*-like detective *Furuhata Ninzaburō* performs a similar fingerman trick, computing $7 \times 8 = 56$ to introduce the episode "Murder of a Mathematician" (1995).

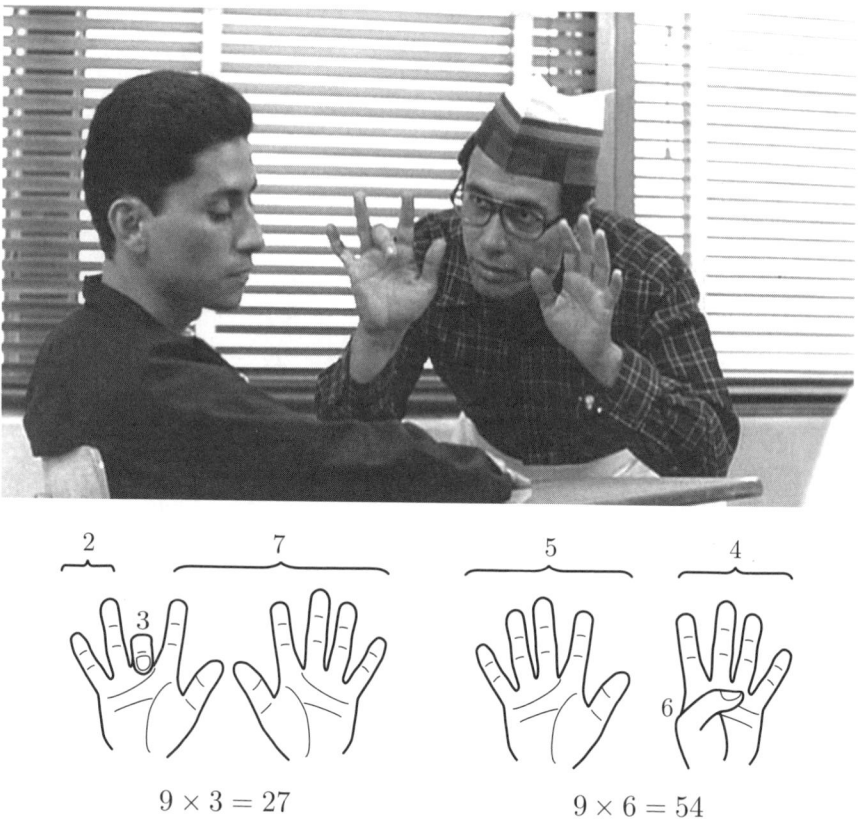

Fig. 3.1 Escalante demonstrates how to multiply by 9.

the Romans were capable of using the concept of zero? It was your ancestors, the Mayas, who first contemplated the zero, the absence of value. True story. You burros have math in your blood.

Escalante is substantially correct. The Mayans, unlike the Greeks and the Romans, had a certain conception of zero, of positional notation for numbers. For example, in the number 23 the 2 means "two tens," and in the number 203, the 2 means "two hundreds": it is the 0 which acts as a placeholder, keeping the distinction clear.

However, having 0 as a placeholder is not the same as understanding 0 to be a number in and of itself, as in the equation $2 - 2 = 0$. The Mayans seemingly did not have this more subtle understanding of 0. Around 800 AD (and prior to the Mayans), a succession of Indian mathematicians struggled, ultimately successfully, with the more precise concept of 0 as a number.[5]

---

[5] For a lovely exposition of the history of zero, see Robert Kaplan, *The Nothing That Is* (Penguin, London, 1999).

Fig. 3.2 Escalante hammers home the multiplication of negative numbers.

After his description of zero, Escalante goes on to discuss parentheses and negative numbers (figure 3.2):

ESCALANTE: Oralé! Okay! Parentheses means multiply. Every time you see this, you multiply. A negative times a negative equals a positive. A negative times a negative equals a positive. Say it! A negative times a negative equals a positive. Say it!
CLASS AND ESCALANTE: A negative times a negative equals a positive.
ESCALANTE: Again!
CLASS AND ESCALANTE: A negative times a negative equals a positive. A negative times a negative equals a positive.
ESCALANTE: I can't hear you!
CLASS AND ESCALANTE: A negative times a negative equals a positive.
ESCALANTE: Louder!
CLASS AND ESCALANTE: A negative times a negative equals a positive.
ESCALANTE: Louder!
CLASS AND ESCALANTE: A negative times a negative equals a positive.
ESCALANTE [softly]: *Why?*

The above scene emphasizes two aspects of the teaching of mathematics: first, the drilling of the fact, until it is accepted as true without question; and second the question: *why* is it true? (And the reader may well ponder, why indeed is a negative times a negative equal to a positive?)

3.3 Let X Be the Number of Girlfriends

The movie continues with a number of scenes of Escalante teaching algebra. The mathematics mostly appears as background illustration, including a great image of Angel as sinner (figure 3.3).

$$x^2 - 17x + 6 = (7x +)$$
$$14\,x^2 - 17x + 6 = (7$$

Fig. 3.3 Angel being sacrificed on the Altar of Polynomials.

After the algebraic introductions, we see a problem discussed in detail.

0:32

CLASS [reading from the blackboard]*: Juan has five times as many girlfriends as Pedro. Carlos has one girlfriend less than Pedro. The total number of girlfriends between them is twenty. How many does each gigolo have?*

ESCALANTE: Anybody. [Tito raises his hand.] *Think you got it, Einstein? You think you gonna do it?*

TITO: Juan is X. Carlos is Y. Pedro is X plus Y. Is Pedro bisexual or what?

ESCALANTE: I have a terrible feeling about you. [Tito blows Escalante a kiss.]

CLAUDIA: Kemo. 5X equals Juan's girlfriends?

ESCALANTE: You're good now, but you're gonna end up barefoot, pregnant, and in the kitcheeen. [Laughter from the class.]

RAFAELA: Can you get negative girlfriends?

ESCALANTE: No, just negative boyfriends. [He looks up to the heavens.] *Please forgive them, for they know not what they dooo!*

ANGEL: Carlos has X minus 5 girlfriends, que no?

ESCALANTE: Que no? is right. Que no. [Lupe raises her hand.] *The answer to my prayers!*

LUPE: May I go to the restroom please? [Laughter.]

ESCALANTE: In ten minutes. Hold it. [He walks up to Javier.] *Señor Maya. Hit it.*

JAVIER [Javier smugly wiggling his pencil]: *It's a trick problem, Mr. Kemo. You can't solve it unless you know how many girlfriends they have in common. Right?*

ESCALANTE: It's not that they're stupid. It's just that they don't know anything.

JAVIER: I'm wrong?

ANA [who has just appeared at the door]: *X equals Pedro's girlfriends, 5X equals Juan's girlfriends, X minus 1 equals Carlos's girlfriends. X plus 5X plus X minus 1 equals 20, so X equals 3.* [The class claps, as Ana sits down.]

Ana gives the expected answer, and the algebra is correct, but notice that Javier is also correct. Escalante is expecting the students to assume that no girl is the friend of more than one gigolo; as Javier has noticed, without this assumption, the problem has more than one solution.

3.4 Newton Was an Idiot

Escalante decides that he will prepare the class for the Advanced Placement mathematics exam (primarily calculus). The parents' permission is required for attendance at the extra classes, and we witness one student cajoling her mother.

0:41

CLAUDIA: Mom, calculus is math that Sir Isaac Newton invented so he could figure out planet orbits, but he never bothered to tell anybody about his discovery until this other scientist guy went around claiming he had invented calculus. But the guy was so stupid that he got it all wrong, and so Newton had to go public and correct his mistakes. Don't you think that's neat?

CLAUDIA'S MOM: For a genius, Newton was an idiot.

Claudia is referring to Gottfried Willhelm Leibniz, the co-inventor of calculus. Claudia's version of the history is correct, except for the implicit suggestion that Leibniz may not have come up with the ideas independently and the explicit claim that Leibniz was stupid and got it all wrong.

We next see the students, who have come to class very early in the morning. They are being taught a standard calculus technique for calculating volumes of revolution, "by disks." So the graph $y = x^2$ from $x = 0$ to $x = 2$ is rotated around the y-axis, creating a bowl. The radius of the bowl at any height y is then x, giving the area πx^2 of a horizontal circular slice through the bowl. This leads to Escalante's formula on the board (figure 3.4),

$$V = \pi \int_0^2 x^2 \, dy \,.$$

Fig. 3.4 Escalante teaches volume by disks.

However, the students are struggling merely to stay awake:

0:44

ESCALANTE: One, you got the graph, right here. Two, the strip, the most important part is right here. It's the radius of rotation. That's it. Anybody got any questions? [Escalante sees Tito, asleep, and Escalante starts slapping him gently with a towel.] *Wake up this morning! How are you?*

TITO [groggily]: *I was swimming with dolphins whispering imaginary numbers, looking for the fourth dimension.*

ESCALANTE [pushing Tito's head down onto the towel]: *Good! Go back to sleep. That's very good.*

In the next math scene, we see Pancho at the blackboard struggling, and failing, with an integration by parts problem (figure 3.5). He is attempting to compute the integral

$$\int x^2 \sin x \, dx = -x^2 \cos x + 2x \sin x + 2 \cos x + C \,.$$

Pancho's calculations are correct but unhelpful, since he has made the wrong choices for u and dv (using the usual symbols).

0:48

ESCALANTE: Try the shortcut. This is easy. Baby stuff for boy scouts.

PANCHO: Kemo, my mind don't work this way!

ESCALANTE: Tic-tac-toe! It's a piece of cake upside down. Watch for the green light. [Pancho stares at the board, and then thumps it with his fist.]

Fig. 3.5 Tito struggles with integration by parts.

PANCHO: I've been with you guys two years! Everybody knows I'm the dumb-est. I can't handle calculus!
ESCALANTE: Do you have the ganas? Do you have the desire?
PANCHO: Yes! I have the ganas!
ESCALANTE: You want me to do it for you!
PANCHO: Yes!
ESCALANTE: You're supposed to say "no"! [Escalante continues with the integration problem, and talks to Pancho as he writes.] *Tic. Tac. Toe. Simple.*

Escalante actually gives a very nice method for performing certain integration by parts problems, where the u term, a power of x, will be differentiated out (see figure 3.6): the first column contains u and its derivatives, until we arrive at 0; the second column contains $\frac{dv}{dx} = \sin x$ and its antiderivatives; and the third column contains alternating positives and negatives. Then, scoring off as indicated, he obtains all the terms produced by the integration.

Notice that by emphasizing his shortcut, Escalante also ignores Pancho's misunderstanding. As with the Finger Man trick for multiplication by 9, this is indicative of Escalante's style (in real life as well as in the movie): he places much more emphasis on drill and techniques than on fundamental understanding. The scene also uses Escalante's magic word: *ganas*. For Escalante, all that matters is *desire*. That is the message of the movie.

Escalante works himself into the ground, to the point of having a minor heart attack. We see this coming when Escalante asks the class about the function $f(x) = 3\sin(2x + \pi)$, but is unable to see what's going wrong (figure 3.7).

$$\int x^2 \sin x \, dx =$$

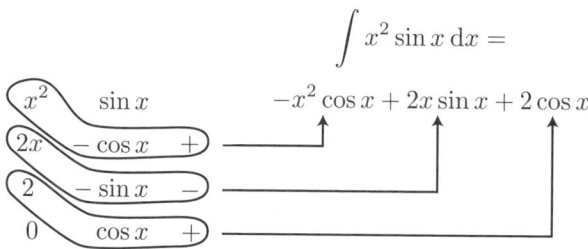

Fig. 3.6 Escalante's shortcut.

0:52

ESCALANTE: We're looking for the area in the first quadrant bounded by the curve. What are the limits? Anybody.

TITO: 0 to $\frac{\pi}{2}$, sir.

ESCALANTE: Wrong. Lupe.

LUPE: 0 to $\frac{\pi}{2}$?

ESCALANTE: What's wrong with you? This is review.

LUPE: Kemo, I checked my work twice.

ESCALANTE: I'm giving you the graph. Check it again.

ANGEL: No, Kemo, I'm getting the same answer as the gordita.

LUPE: Don't call me gordita, pendejo!

ANA: It's 0 to $\frac{\pi}{2}$, sir.

JAVIER: Yeah. I got the same thing.

ESCALANTE: You should know this. No, no way. You should know this. What's wrong with you? This is review! You're acting like a blind man in

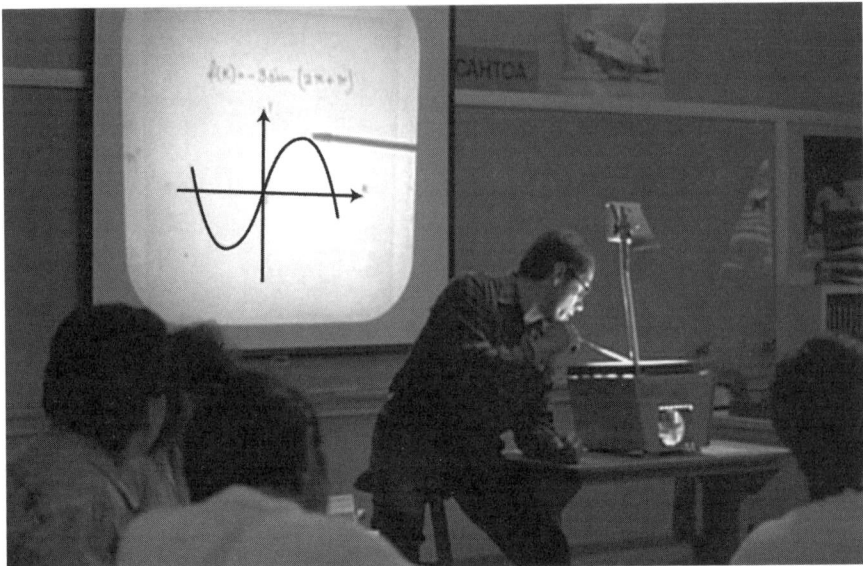

Fig. 3.7 Escalante misleads the students.

a dark room looking for a black cat that isn't there. What's wrong with you guys? I don't believe it. You're giving me a shot from the back! No way! No way!

The problem is that Escalante has misdrawn the graph: the function $f(x) = 3\sin(2x + \pi)$ equals 0 at $x = 0$ and $x = \frac{\pi}{2}$, but the graph is *below* the x-axis (i.e., in the fourth quadrant) between these limits. Thus, in contradiction to Escalante's graph, the correct limits for the first region in the first quadrant are $\frac{\pi}{2}$ to π.

Escalante's blind man description of the students is actually a famous quotation by Charles Darwin, describing *all* mathematicians and the very nature of their pursuit. Escalante's use of it as a description of the students (supposedly) making simple mistakes is quite different.

3.5 The Students Stand and Deliver, Again

Finally, the students take the Advanced Placement exam. We see them struggle with the questions, and the relief afterward. But then comes the central drama of the movie, when the students are accused of cheating by the external authorities.

Reluctantly, the students retake the exam. In the climactic scene, we hear the students' scores being read out over the phone: they've all passed, and the movie ends with Escalante walking down the school corridor, pumping his fist in victory.

3.6 Will the Real Jaime Escalante Please Stand Up?

As we remarked above, the story of *Stand and Deliver* is by and large true. The major fictional aspect of the movie was to compress many years of work, with many classes of students, into two years with one class.

Jaime Escalante started at Garfield High in 1974. The school already had an Advanced Placement math program, but it was in disrepair. In 1978 Escalante went to work on it. The first such class started with fourteen students; five students lasted to take the exam in 1979, of whom two students passed.

Each year, the size and the success of Escalante's program grew. *Stand and Deliver* tells the story of the class of 1981–1982, when eighteen students took the exam. Fourteen of the students were (implicitly) accused of cheating; all those who were not accused passed the exam. The accused students were offered an alternate exam; two students declined, having already received university offers. The other twelve accused students sat the alternate exam, and all twelve passed. As the movie indicates at the very end, Escalante's program then continued to grow, becoming arguably the most successful such program in the country.

Although the compression of time leads to a dramatically powerful movie, it is also misleading in an important sense. The movie overstates how quickly Escalante could turn around the mathematics program, and how quickly he could take a student from mathematical ignorance to recognized success. This also goes some way to explaining why Escalante's classes continued to do better, year after year.

Another element of confusion in the story concerns Escalante's goal. Through teaching mathematics, Escalante attempted to make his students appreciate their own abilities and possibilities. The goal was not primarily to have the students understand and appreciate the mathematics; the goal was to pass the exam, which was the meaningful and practical achievement for the students. This part of the story is somewhat underemphasized in the movie.

The quick-trick approaches Escalante instilled would also tend to lead to the students making similar and peculiar mistakes, and goes part of the way to explain the suspicion triggered by the students' performance. However, there is another reason why the students' performance triggered suspicion: their performance was in fact suspicious. In reality it is not clear whether some of the students cheated or not (though by passing the test they proved that they didn't need to). This is discussed in some detail in Jay Matthews' excellent book; see footnote 2 on page 41.

Finally, there is one last manner in which the movie is misleading. In *Stand and Deliver*, Escalante is presented as a strong-willed and somewhat acerbic character, but the movie gives no real sense of the extent to which Escalante was willing and able to really piss people off. In real life, Escalante did not suffer fools, and at Garfield High, Escalante's pushy, critical arrogance made him plenty of enemies. Effectively forced out, Escalante left Garfield High in 1991. The Advanced Placement program at Garfield then quickly crumbled.

But these are the most minor of quibbles. *Stand and Deliver* is a wonderful movie about an incredible story, and a truly great teacher (figure 3.8).

Fig. 3.8 Detail of a mural entitled Los Angeles Teachers by Hector Ponce, showing Jaime Escalante together with Edward James Olmos. The mural is located in Los Angeles at the corner of Wilshire Boulevard and South Alvarado Street, facing MacArthur Park.

Chapter 4
The Annotated Pi Files

Darren Aronosky's π (1998) is an excellent, dark movie. Featuring a clever script, very good acting, and plenty of beautiful mathematics, this is a must-see cult movie, and a great source for mathematical movie clips.

The story is of the brilliant mathematician Max Cohen, played by Sean Gullette. Max is struggling to discern a pattern behind the numbers that make up the stock market. Along the way, he comes across some of mathematics' icons: π, the Fibonacci numbers, and the golden ratio.

Max also gets mixed up with the numerology of Jewish mysticism. In fact, the answer to Max's question seems to be hidden in a 216-digit number that his new religious friend believes to be the name of God. The number seems to be the key to everything, with the power to wreak havoc: Max's computer Euclid repeatedly crashes while analyzing patterns that supposedly conceal the number; Max suffers debilitating migraine attacks; and Sol Robeson, Max's former PhD advisor, suffers two strokes, the second leading to his death. Finally, Max resorts to drilling into his brain, to eradicate all traces of the number. (Well, we told you the movie was dark.)

Here we give an annotated version of the story, told through the mathematical scenes and screenshots.

4.1 Max the Mathematician

0:02
This scene introduces Max as someone who is brilliant with numbers. As Max leaves his apartment, the excited young Jenna runs up to him.

JENNA: Max! Max! Can we do it?
MAX: Jenna.
JENNA: Three hundred and twenty-two times four hundred and eighty-one.
[Jenna types it into her pocket calculator.]
MAX [instantly]*: One hundred fifty-eight thousand, one hundred two. Right?*
JENNA: Right!

Actually there is a mistake here. Jenna should have said 322×491 instead of 322×481, according to the director's shooting script, and in order to get the answer in the movie.[1]

Max heads down the staircase and Jenna screams after him.

JENNA: Okay, seventy-three divided by twenty-two.
MAX [instantly again]*: Three point three one eight one eight one eight ...* $[\frac{73}{22} = 3.3181818\ldots]$.

Here, Max's voice trails off as he walks down the stairs, "one" on one step and "eight" on the next. It is a poetic portrayal of the infinite decimal expansion of this number.

4.2 Mathematics Is the Language of Nature

0:03
MAX [voiceover]*: Restate my assumptions:*

1. *Mathematics is the language of nature.*
2. *Everything around us can be represented and understood through numbers.*
3. *If you graph the numbers of any system, patterns emerge.*

Therefore there are patterns everywhere in nature. Evidence: the cycling of disease epidemics, the wax and wane of Caribou populations, sunspot cycles, the rise and fall of the Nile. So what about the stock market? The universe of numbers that represents the global economy. Millions of human hands at work. Billions of minds, a vast network screaming with life, an organism, a natural organism. My hypothesis: Within the stock market there is a pattern as well, right in front of me, hiding behind the numbers. Always has been.

This elegantly summarizes Max's approach to his research which, according to him, is number-theoretic in nature. Later, it degenerates into (what is in the real world) numerology.

Sixteen minutes into the commentary, director Darren Aronofsky remarks upon the math in the movie: "So, all this math stuff is real stuff. Fibonacci is a real dude or was a real dude and the golden spiral was real, we did not make it up for the film. It is just a lot of plagiarism from the Bible, from math text books, from Pythagoras, we kind of stuck it all together."

4.3 Pattern in Pi

0:10
Max is playing Go with Sol, his former advisor.

---

[1] The director's shooting script includes deleted scenes, containing more math. At the time of writing, this script was freely available from various websites. Highly recommended reading.

SOL: You haven't taken a single break.
MAX: I'm so close.
SOL: Have you met the new fish my niece bought me? I named her Icarus. After you, my renegade pupil. You fly too high, you'll get burned. The more I see you, the more I see myself thirty years ago. My greatest pupil published at sixteen, PhD at twenty. But life isn't just mathematics, Max. I spent forty years looking for patterns in π. I found nothing.
MAX: You found things.
SOL: I found things, but not a pattern.

It is clear that by "pattern," Max and Sol mean some sort of pattern in the decimal expansion 3.1415... of π. In fact, investigation of the decimal expansion of π is a comparatively meaningless endeavor for a top-notch mathematician.

There are *numbers* and there are the different *ways of writing/representing numbers*, and it is easy for nonmathematicians to confuse the two. The (base ten) decimal expansion of a number is a useful convention, but base ten is used simply because of the biological accident that (most) humans have ten fingers. However, writing a number in base ten seldom provides any special or deep insight into the number.

For example, at least theoretically we know whether or not a number is rational by looking at its full decimal expansion. A number is rational if (and only if) its decimal expansion winds up repeating forever, just like Jenna's $\frac{77}{22} = 3.318181818...$ above. However, this is just as true if we expand the number with respect to any other base, such as base 2 or base 3.

As an illustration, π is known to be an irrational number (although this was not discovered by examining its decimal expansion). So both its well known decimal expansion and its binary expansion

$$11.001001000011111101101010101000010000\ldots$$

go on forever, without repeating.[2]

While it is important to realize that the decimal expansion of π plays only a minor role in the study of this important number, it is equally important to realize that mathematics is not as dehumanized a subject as it is often portrayed. Entirely human motives and aesthetics are often responsible for the obsession of mathematicians with officially unimportant topics. The decimal expansion of π is one such example, regularly providing challenges for new supercomputers. And some very good mathematicians spend considerable time

---

[2] Are you thinking of memorizing a couple of hundred digits of the decimal expansion of π but are unsure where to stop? If so, we recommend memorization to the 767th decimal place. If you do this, you will be able to finish reciting your expansion proudly and correctly with "... nine, nine, nine, nine, nine, nine, and so on." These six consecutive 9s are sometimes called the Feynman point, named after the famous physicist Richard Feynman, who once remarked on the humorous "application" of this curiosity.

puzzling over problems related to this fairly random choice of writing π, as well as the more general question of writing π with respect to other bases.

Given this general obsession with the decimal expansion of π, it may come as a surprise that mathematicians know very little about the patterns of this most popular way of writing π. While it is simple to compute larger and larger portions of the decimal expansion of π, we do not even know whether each of the different digits 0 to 9 occurs infinitely often. Still, there have been major advances in our understanding of these expansions. Perhaps the most remarkable is the discovery of a formula enabling us to calculate any desired digit in the binary expansion of π, without having to calculate any of the preceding digits.[3]

Finally, if Max is deeply concerned with the decimals of π, Darren Aronofsky is clearly much less so. The background for the title sequence of the movie is thousands of digits of the decimal expansion of π, but only the first nine digits are correct (figure 4.1). Looking closely at the sequence of numbers rolling across the screen, you notice giveaway strings such as 0123456789 popping up. It quickly becomes clear that someone simply made up most of the numbers. It's all a bit surprising: since it is incredibly easy to get hold of the correct decimal expansion of π up to billions of decimal places, it seems it would be harder to get things wrong here than to get them right.[4]

Other mathematical images in the title sequence of the movie are supposed to get us into Max's head during one of his migraine attacks (Möbius strips, spirals, 3D plots of functions, and so on.)

0:12
Max sits in a New York City subway car.

MAX [voiceover]: *Not a pattern—Sol died a little when he stopped research on π. It wasn't just a stroke. He stopped caring. How could he stop when he was so close to seeing π for what it really is? How could you stop believing that there is a pattern, an ordered shape behind those numbers when you are so close? We see the simplicity of the circle, we see the maddening complexity of the endless string of numbers, three point one four, off to infinity.*

While saying this, Max draws a circle and then writes the formulas $A = \pi r^2$ and $C = 2\pi r$. Next he writes $\pi = 3.14159\ldots$[5]

By the way, π is not the first movie in which a message from God (or similar) is encoded in numbers. In the unintentionally hilarious *Red Planet Mars* (1952), scientists employ the decimal expansion of π to communicate with the "Martians." Also, in his science fiction novel *Contact*, Carl Sagan

---

[3] D. H. Baily, P. B. Borwein, and S. Plouffe, On the rapid computation of various polylogarithmic constants, *Mathematics of Computation* 66 (1997), 903–913.

[4] The movie π is definitely not the only one that gets it wrong. Just about every movie character who tries to recite or display more than a few digits of π seems to be doomed to fail; see chapter 18 ("Money-Back Bloopers") for details.

[5] A bit of trivia: The first song on the soundtrack (written by Clint Mansell) has the title "πr^2," and the last song is "$2\pi r$."

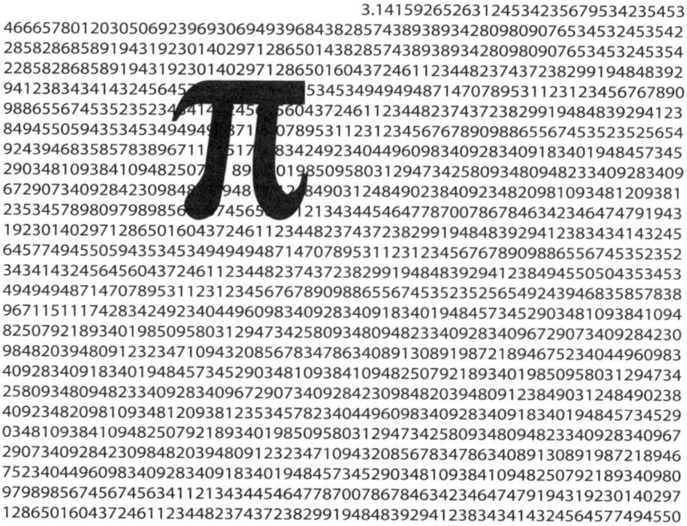

3.14159265263124534423567953423545 3
46665780120305069239693069493968438285743893893428098090765345324535 42
2858286858591943192301402971286501438285743893893428098090765345324535 4
22858286858591943192301402971286501604372461123448237437238299194848392
941238343414324564 5 5345349494949487147078953112312345676789 0
988655674535235234 414 45 5604372461123448237437238299194848 39294123
849455059435345349494 871 07895311231234567678909886556745352352 5654
924394683585783896711 517 8342492340449609834092834091834019484573 45
290348109384109482507 89 0198509580312947342580934809482334092834 09
67290734092842309848 948 7 849031248490238409234820981093481209 381
23534578980979898560 74565 1 21343445464778700786784634234647479194 3
192301402971286501604372461123448237437238299194848392941238343414324 5
645774945505943534534949494871470789531123123456767890988655674535235 2
343414324564560437246112344823743723829919484839294123849455050435345 3
494949487147078953112312345676789098865567453523525654924394683585783 8
9671151117428342492340449609834092834091834019484573452903481093841094
82507921893401985095803129473425809348094823340928340967290734092842 30
98482039480912323471094320856783478634089130891987218946752340449609 83
4092834091834019484573452903481093841094825079218934019850958031294734
258093480948233409283409672907340928423098482039480912384903124849023 8
40923482098109348120938123534578234044960983409283409183401948457345 29
0348109384109482507921893401985095803129473425809348094823340928340967
290734092842309848203948091232347109432085678347863408913089198721894 6
75234044960983409283409183401948457345290348109384109482507921893409 80
97989856745674563411213434454647787007867846342346474791943192301402 97
12865016043724611234482374372382991948483929412383434143245645774945 50

Fig. 4.1 Part of the title sequence of the movie π.

ponders the possibility of finding a message from the creators of the Universe embedded in the base eleven expansion of π. Sadly, this part of the story did not make it into the movie of the same title. Our consolation prize is Jodie Foster discussing prime numbers.[6]

4.4 Numerology: Father + Mother = Child

0:13
While scribbling on paper in a cafe, Max meets Lenny Meyer, a Hasidic Jew.

LENNY: So, what do you do?
MAX: Um, I work with computers, math.
LENNY: Math? What type of math?
MAX: Number theory. Research, mostly.
LENNY: No way, I work with numbers myself. I mean not traditional, though I work with the Torah. Amazing. You know Hebrew is all math, it's all numbers. You know that? Here look. The ancient Jews used Hebrew as their numerical system. Each letter is a number. Like the Hebrew A, Aleph is one. B, Bet, is two, understand? Look at this. The numbers are interrelated. Like take the Hebrew word for father Ab, Aleph, Bet, one [plus] two equals three, all right. Hebrew word for mother Am, Aleph, Mam, one [plus] forty equals forty-one. Sum of three and forty-one, forty-four, right? Hebrew word for child, mother, father, child, (ye)(le)d, that's ten [plus] thirty and four, forty-

---

[6] Which she peculiarly refers to as "base ten numbers," whatever that means.

four. The Torah is just a long string of numbers. Some say that it is a code sent to us from God.
MAX: That's kind of interesting.
LENNY: Yeah, that's just kids' stuff. Check this out. The word for the Garden of Eden, Kadem. Numerical translation one forty-four.

At this point, Lenny scribbles the numbers 4, 30, and 10, which do *not* sum to 144. Referring to the shooting script, we find that the intended numbers are Kuf = 100, Daled = 4, and Mem = 40. Of course, what all this also demonstrates is that the type of nonsense behind The Bible Code is not at all new.[7]

LENNY: Now the value of Tree of Knowledge in the garden, right, Aat Ha Haim, two hundred thirty-three. One forty-four, two thirty-three. Now you can take those numbers.

4.5 The Fibonacci Numbers and the Golden Ratio

MAX: Those are Fibonacci numbers.
LENNY: Huh?
MAX: You know, the Fibonacci sequence.
LENNY: Fibonacci?
MAX: Fibonacci is an Italian mathematician in the thirteenth century.

Here, Max begins writing out the Fibonacci sequence: $1, 1, 2, 3, 5, 8, \ldots$ Remember that in the Fibonacci sequence a new number is obtained by adding the two preceding numbers: $1 + 1 = 2, 1 + 2 = 3, 2 + 3 = 5, 3 + 5 = 8$, and so on. This means that the sequence continues as

$$1, 1, 2, 3, 5, 8, 13, 21, 34, 55, 89, 144, 233, \ldots$$

So 144 and 233 are really Fibonacci numbers, as Max spotted straight away.

MAX: If you divide a hundred and forty-four into two hundred and thirty-three, the result approaches theta.
LENNY: Theta?
MAX: Theta, the Greek symbol for the golden ratio, the golden spiral.

Time to recall some basic facts about the golden ratio θ (theta), or ϕ (phi), as it is now more commonly called. A rectangle is *golden* if, on cutting away a square as pictured in figure 4.2, we are left with a new, rotated rectangle of the same shape. Then the golden ratio is the ratio of the long to the short side of a golden rectangle.

---

[7] Is there anybody who still takes The Bible Code seriously? If so, they should be directed to the very funny and thoroughly debunking website of the Australian computer scientist Brendan McKay. There, they can learn of the hidden messages in *Moby Dick*.

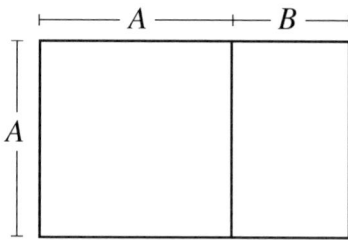

Fig. 4.2 Removal of a square from a golden rectangle leaves a smaller golden rectangle.

Using this geometric definition it is easy to calculate that

$$\phi = \frac{1 + \sqrt{5}}{2} = 1.61803398\ldots\,[8]$$

As Max indicates, the Fibonacci numbers and the golden ratio are closely related. Max mentions one of the most beautiful links, first discovered by Johannes Kepler (of planetary fame): If F_n denotes the nth Fibonacci number, then the ratio of consecutive Fibonacci numbers $\frac{F_{n+1}}{F_n}$ approaches the golden ratio as n goes to infinity:

$$\frac{F_{n+1}}{F_n} \longrightarrow \phi.$$

In fact, this way of approximating the golden ratio gives very good results even for small n. For example, $\frac{F_{13}}{F_{12}} = \frac{233}{144} = 1.6180555\ldots$ approximates the golden ratio correctly to four decimal places.[9]

The golden ratio and the Fibonacci numbers appear in many natural phenomena, which is why Max is so interested in them.

MAX: Theta, the Greek symbol for the golden ratio, the golden spiral.

---

[8] At a later point, Max begins this calculation. He calls the short side of the golden rectangle A and the long side $A + B$. Then the defining relation of a golden rectangle translates into the equation

$$\frac{A + B}{A} = \frac{A}{B},$$

or equivalently

$$1 + \frac{1}{\frac{A}{B}} = \frac{A}{B}.$$

After substituting $\phi = \frac{A}{B}$ and multiplying both sides by ϕ, this equation turns into a quadratic equation. Doing the familiar $b^2 - 4ac$ stuff, we find that the positive solution to this equation is $\frac{1+\sqrt{5}}{2}$.

[9] However, this particular link is weaker than first appears. Replace the two 1s at the beginning of the Fibonacci sequence by your two favorite positive numbers, and generate a sequence of numbers using the same Fibonacci addition rule. Then the ratio of consecutive numbers in this new sequence will also approach the golden ratio.

Max draws what is probably supposed to be a *logarithmic spiral*: this is a special spiral with the property that the ratio of two successive radii does not depend upon the direction. This fixed ratio, which can be any positive number, determines the precise proportions of the spiral. Logarithmic spirals approximate many naturally occurring spirals, such as the one apparent in the cross section of a nautilus shell (figure 4.3).

Fig. 4.3 A logarithmic spiral in a nautilus shell, with two consecutive radii highlighted.

LENNY: Wow, I never saw that before. That's like that series that you find in nature, like the face of a sunflower?
MAX: Wherever there's spirals.

Here, Max ponders a spiral created by dropping a dollop of cream into his coffee.

LENNY: See, there's math everywhere.

What do spirals, and, in particular, logarithmic spirals have to do with the golden ratio and the Fibonacci numbers? A bit, but not as much as Max suggests, or nearly as much as is commonly believed.

A first and very striking link, with which Lenny seems to be familiar, relates to the roughly logarithmic spirals visible in many flower heads: count the numbers of left-winding and right-winding spirals, and you usually get two consecutive Fibonacci numbers. In sunflowers these two numbers are, depending on the size of the flower head, 21 and 34, 34 and 55, 55 and 89, or 89 and 144. The presence of these Fibonacci numbers can be explained in terms of mathematical properties of the golden ratio.[10]

There are many other genuine occurrences of these numbers in nature. Regrettably, Max picks one of the weakest examples as one of his guiding

---

[10] For a well-written and accessible account of these ideas, see Mario Livio's excellent book *The Golden Ratio* (Broadway, New York, 2003).

principles: the commonly touted links between many of the logarithmic spirals occurring in nature and the golden ratio are mostly the product of wishful thinking and screwy logic.

True, there is indeed a natural logarithmic spiral associated with the golden ratio, and Max actually constructs it later on in the movie. Rather than wait for Max, let's have a closer look now.

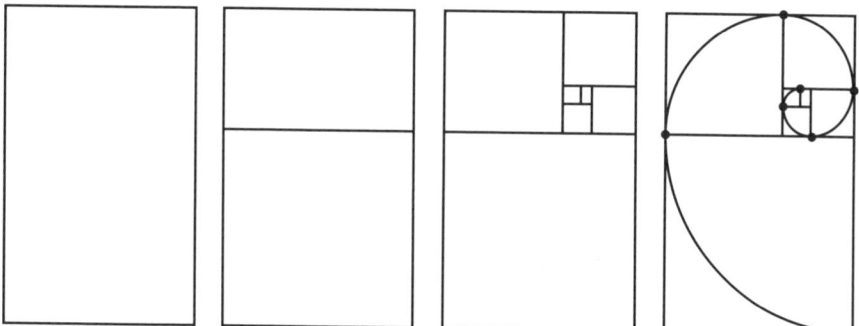

Fig. 4.4 Constructing a spiral from a golden rectangle.

Cut away a square from a golden rectangle, as in the second diagram of figure 4.4. Then you are left with a new, smaller, golden rectangle. Cut a second square from this new golden rectangle and you get another even smaller golden rectangle. Repeating, and highlighting the squares that are cut off, we obtain the third diagram above. Next, inscribe a quarter circle in each of the squares, as in the fourth diagram, and the result is a spiral.

Now, it is often claimed that this spiral is logarithmic. This is clearly nonsense, since logarithmic spirals don't contain circle segments. However, there is a specific logarithmic spiral that contains all the indicated (infinitely many) corners of the squares. And this logarithmic spiral does look very similar to the spiral composed of quarter circles.

Okay, so there is a logarithmic spiral associated with the golden ratio in a fairly natural manner. Its determining ratio of consecutive radii is $\phi^4 = 6.854\ldots$ However, as we remarked above, *every* positive real number has an associated logarithmic spiral, and all these spirals are different in shape.

Therefore, it makes no sense to claim, as is so often done, that every logarithmic spiral we find in nature must automatically be the spiral associated with the golden ratio. For example, in the cross section of a nautilus shell, we can see a spiral which is definitely close to being logarithmic. However, contrary to common belief this spiral is usually not even close to the golden ratio spiral.[11]

---

[11] C. Falbo, The golden ratio—A contrary viewpoint, *College Mathematical Journal* 36 (2005), 123–134.

Later in this chapter, and also in chapter 5 ("Nitpicking in Mathmagic Land"), we will continue our discussion of true and imagined properties of the golden ratio and the Fibonacci numbers.[12] For a list of other movies that feature these numbers see section 21.7.

Back to the story.

0:16

MAX [voiceover]: *Restate my assumptions:*

1. *Mathematics is the language of nature.*
2. *Everything around us can be represented and understood through numbers.*
3. *If you graph the numbers of any system, patterns emerge. Therefore there are patterns everywhere in nature.*

As Max says this, he draws a square spiral whose vertices seem to coincide with certain entries of the stock market charts he is studying; see figure 4.5.[13]

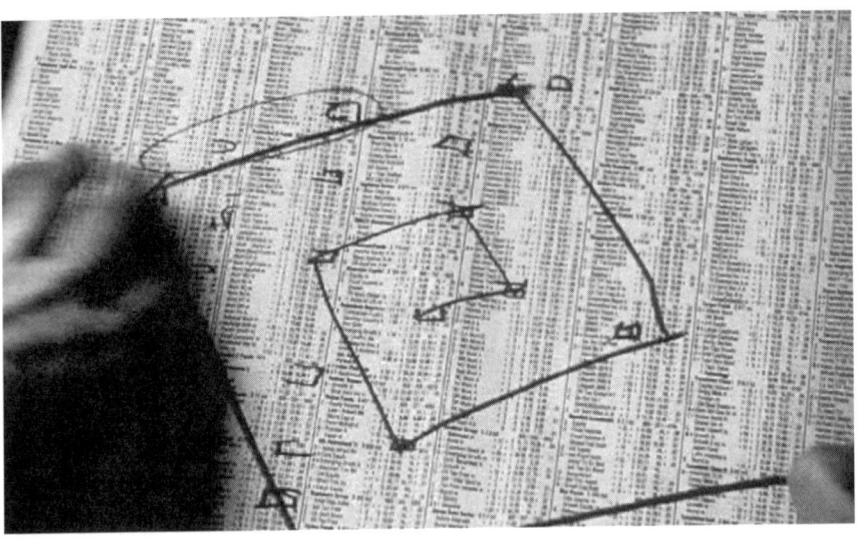

Fig. 4.5 "if you graph the numbers of any system, patterns emerge."

MAX [voiceover]: *So what about the stock market? The universe of numbers that represents the global economy. Millions of human hands at work. Billions of minds, a vast network screaming with life, an organism, a natural organism. My hypothesis: Within the stock market there is a pattern, right in front of me, playing with the numbers. Always has been.*

---

[12] For an excellent account separating fact from fiction, see Mario Livio's book referred to above in footnote 10.

[13] A similar spiral drawing appears in *The Giant Claw* (1957), when the heroes try to determine the path of the chicken monster.

Clearly, Max believes he has had a major insight. Following up on this he
runs a program on Euclid, his computer, to predict the stock market. The
results seem absurd, and Euclid crashes. However, before it crashes Euclid
prints out a long string of seemingly random numbers. At the same time Max
passes out from one of his migraines.

4.6 Archimedes the Goldfish

0:20
Sol's apartment. Max and Sol are playing Go.

MAX: Euclid crashed. I lost all my data, my hardware.
SOL: Your mainframe?
MAX: Burnt.
SOL: What happened?
*MAX: First I get these crazy low picks. Then Euclid spits out this long string
of numbers. Never saw anything like it and then it fries. The whole machine
just crashed.*
SOL: You have a printout?
MAX: Of what?
SOL: Of the picks, the number?
MAX: I threw it out.
SOL: What was the number it spit out?
MAX: I don't know, just a long string of digits.
SOL: How many?
MAX: I don't know.
SOL: What was it, a hundred, a thousand, two hundred sixteen [this is im-
portant!] *How many?*
MAX: I don't know. Probably around two hundred. Why?
*SOL: I dealt with some bugs back in my π days. I was wondering if it was
similar to one I ran into. Have you met Archimedes* [one of his fish]*? The
one with the black spots. You see?*
MAX: Yeah.
*SOL: You remember Archimedes of Syracuse? The King asks Archimedes to
determine if a present he's received is actually solid gold. Unsolved problem
at the time. It tortures the great Greek mathematician for weeks. Insomnia
haunts him and he twists and turns in his bed for nights on end. Finally, his
equally exhausted wife, she's forced to share a bed with this genius, convinces
him to take a bath, to relax. While he is entering the tub Archimedes notices
the bathwater rise. Displacement. A way to determine volume. And thus, a
way to determine density, weight over volume. And thus, Archimedes solves
the problem. He screams "Eureka!" and is so overwhelmed he runs dripping
naked through the streets to the King's palace to report his discovery. Now,
what is the moral of the story?*
MAX: That a breakthrough will come.

SOL: Wrong. The point of the story is the wife. You listen to your wife, she will give you perspective. Meaning, you need a break, you have to take a bath, or you'll get nowhere. There will be no order, only chaos. Go home, Max, and you take a bath.

The credits of the movie include consultants for Go, Judaica, and medicine, but none for mathematics. From Aronofsky's commentary we know that his father is a scientist. The story about Archimedes in the movie was told to the director by his father. In his commentary, the actor Sean Gullette remarks that all the math material that we see in the movie was compiled by the director.

Sol telling the story of Archimedes and the gold crown is one of the best movie clips about Archimedes. For other appearances of Archimedes, see chapter 20.

4.7 Coincidence, or Is It?

0:26
In a synagogue.

LENNY: You know when you told me that you were Max Cohen, I didn't realize that you were the Max Cohen. Your work is revolutionary, you know that. It's inspired the work that we do.
MAX: It has?
LENNY: Yes, very much so. The only difference is, we're not looking at the stock market. We're searching for a pattern in the Torah.
MAX: What kind of pattern?
LENNY: We're not sure. We only know, it's 216 digits long.

0:28
Back in Sol's apartment.

MAX: What's the two hundred and sixteen number, Sol?
SOL: Excuse me?
MAX: You asked me if I had seen a two hundred and sixteen digit number, right?
SOL: Oh, you mean the bug. I ran into it working on π.
MAX: What do you mean "ran into it"?
SOL: What is this all about?
MAX: There's these religious Jews that I've been talking to.
SOL: Religious Jews?
MAX: Yeah, you know Hasids, the guys with the beards. I met one in the coffee shop. It turns out the guy's a number theorist. The Torah is his data set. He tells me that they are looking for a two hundred and sixteen digit number in the Torah.
SOL: Come on, it's just a coincidence.
MAX: There's something else, though.

SOL: What?

MAX: You remember those weird stock picks I got.

SOL: Yesterday's stock picks, yes?

MAX: It turns out they were correct. I got two picks on the nose. Smack on the nose Sol. Something's going on. It has to do with that number. There is an answer in that number.

SOL: Come with me.

Sol and Max are playing Go again.

SOL: The Ancient Japanese considered the Go board to be a microcosm of the Universe. Although, when it is empty it appears to be simple and ordered, the possibilities of game play are endless. They say that no two Go games have ever been alike. Just like snowflakes. So, the Go board actually represents an extremely complex and chaotic universe and that's the truth of our world, Max. It can't be easily summed up with math. There is no simple pattern.

MAX: But as a Go game progresses, the possibilities become smaller and smaller. The board does take on order. Soon, all moves are predictable.

SOL: So, so?

MAX: So, maybe, even though we're not sophisticated enough to be aware of it, there is a pattern, an order, underlying every Go game. Maybe that pattern is like the pattern in the stock market, the Torah. This two sixteen number.

SOL: This is insanity, Max.

MAX: Or maybe it's genius. I have to get that number.

SOL: Hold on, you have to slow down. You're losing it, you have to take a breath. Listen to yourself. You're connecting a computer bug I had, with a computer bug you might have had, and some religious hogwash. If you want to find the number two sixteen in the world, you'll be able to find it everywhere. Two hundred sixteen steps from your street corner to your front door. Two hundred sixteen seconds you spend riding on the elevator. When your mind becomes obsessed with anything, you will filter everything else out and find that thing everywhere. Three hundred and twenty, four hundred and fifty, twenty-three. Whatever! You've chosen two hundred sixteen and you'll find it everywhere in nature. But Max, as soon as you discard scientific rigor, you are no longer a mathematician. You are a numerologist.

Great speech by Sol here: if only Max had listened! Who has not noticed a certain number, and then seen this number appear again and again? Of course, on closer inspection most of these "unlikely" events will be found to be quite likely. For example, in a room with 23 or more people, there is a greater than 50:50 chance that two of the people will share a birthday. It feels wrong, until you do the simple math.[14]

---

[14] For an excellent movie about one man's obsession with one particular number, watch Jim Carrey's *The Number 23* (2007). A lot of fun, but no math, except for the number 23 appearing over and over and over and over.

4.8 Back to the Golden Ratio

0:41
In his apartment, after Max has been pondering yet more spirals.

MAX: Remember Pythagoras. Mathematician, cult leader, Athens, circa 500 BCE. Major belief: The Universe is made of numbers. Major contribution: The golden ratio. Best represented geometrically as the golden rectangle. Visually there exists a graceful equilibrium between the shape's length and width. When it is squared, it leaves a smaller golden rectangle behind, with the same unique ratio. The squaring can continue smaller and smaller and smaller, to infinity.

As Max says this, he draws the golden spiral as we described it earlier; compare figures 4.4 and 4.6.

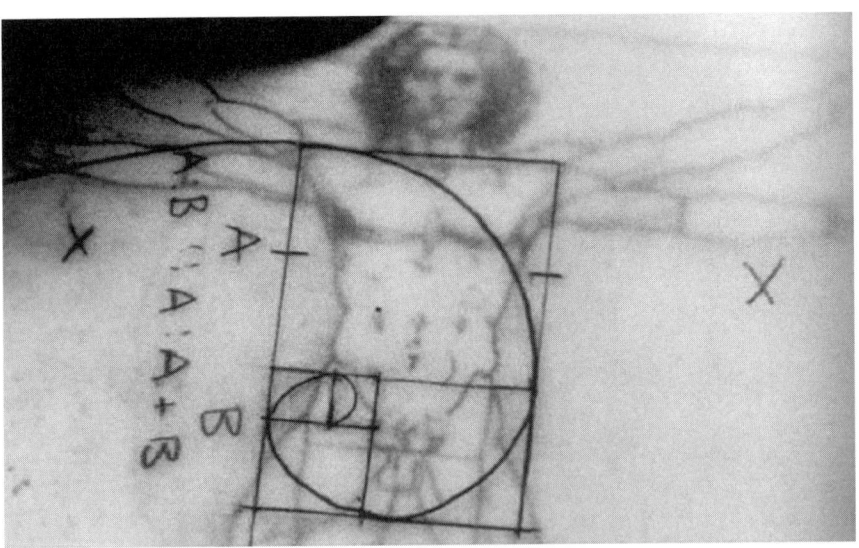

Fig. 4.6 Max superimposes the golden spiral on DaVinci's *Vitruvian Man.*

Max makes a number of mistakes here. First, crediting Pythagoras with the discovery of the golden ratio is suspect; see chapter 5 ("Nitpicking in Mathmagic Land") for details. Second, he reiterates the popular myth that golden rectangles are aesthetically ("perfectly") proportioned. In fact, there is *no* reliable evidence that winners of Rectangle Beauty Contests have close-to-golden proportions. The reason that Leonardo DaVinci's drawing shows up here is because it is supposed to feature perfect golden proportions, which is not the case. We'll have much more to say about this in chapter 5. Third, using the expression "squaring" to mean "removing a square" is peculiar.

Finally, and most importantly, the comparison of ratios should actually be written

$$A : B :: A + B : A.$$

For all our criticism, Max gets it substantially correct (not only in terms of the spiral).

0:42

MAX [voiceover]*: More evidence: Remember Da Vinci, artist, inventor, sculptor, naturalist, scientist, Italy fifteenth century, rediscovered the balanced perfection of the golden rectangle and penciled it into his masterpieces. Connecting a curve through the concentric golden rectangles, you generate the mythical golden spiral. Pythagoras loved this shape, for he found it everywhere in nature: the nautilus shell, rams horns, whirlpools, tornados, our fingerprints, our DNA, and even our Milky Way.*

The word "concentric" makes no sense in this context. Indeed, all that Max says here is basically nonsense.

4.9 Staring into the Sun

0:43

Walking around in Chinatown.

MAX [voiceover]*: When I was a little kid my mom told me not to stare into the sun. So, once when I was six I did. At first the brightness was overwhelming but I had seen that before. I kept looking, forcing myself not to blink and then the brightness began to dissolve. My pupils shrank to pinholes and everything came into focus and for a moment I understood. My new hypothesis: If we are built from spirals, living in a giant spiral, then everything we put our hands to is infused with the spiral.*

This is reminiscent of the famous story about the Belgian physicist Joseph Plateau. Plateau, master of soap films, is well known among mathematicians and bubbleologists for what are now known as Plateau's Problem and Plateau's Rules.[15] Plateau is also famous for being stupid enough to perform Max's exact experiment and find out what happened if he stared at the Sun. And he indeed found out: Plateau went blind.

Max makes a deal with some stock market analysts who want to get their hands on his results on predicting the stock market. This gets him a new powerful computer chip. He also offers to help Lenny to find the number and

---

[15] Plateau's Problem: After a wire loop is dipped in soap solution, what is the shape of the soap film that forms? Plateau's Rules: 1. Soap films attach to smooth surfaces at right angles. 2. In a soap bubble cluster, soap films always join along an edge in threes, and the angle between any two such films is 120 degrees. 3. Where edges meet at a point, they always do so in fours, and at the same angle as the four segments joining the center with the four vertices of a regular tetrahedron.

gets the Torah on a disk from him. Max starts analyzing the Torah with his fixed and chip-enhanced computer Euclid. Euclid crashes, and so does Max (another violent migraine).

0:52

Euclid's monitor is filled again with the mysterious string of numbers that Max encountered previously. This time we get a complete view:

$$94143243431512659321054872390486828512913474876027$$

$$67195923460238582958304725016523252592969257276553$$

$$64363462727184012012643147546329450127847264841075$$

$$62234789626728592858295347502772262646456217613984$$

$$829519475412398501.$$

On closer inspection, it turns out that there are actually 218 digits and not 216 as we were led to believe earlier. As usual in the movie business, near enough is good enough.

0:57

Sol's apartment.

MAX: You lied to me.

SOL: Okay, sit down. I gave up before I pinpointed it. But my guess is that certain problems cause computers to get stuck in a particular loop. The loop leads to meltdown, but just before they crash they—they become "aware" of their own structure. The computer has a sense of its own silicon nature and it prints out its ingredients.

MAX: The computer becomes conscious?

SOL: In some ways—I guess.

MAX [to himself]: Studying the pattern made Euclid conscious of itself. Before it died it spit out the number. That consciousness is the number.

SOL: No, Max, it's only a nasty bug.

MAX: It's more than that!

SOL: No it's not! It's a dead end, there is nothing there.

MAX: It's a door Sol, a door.

SOL: A door in front of a cliff. You're driving yourself over the edge. You need to stop.

MAX: You were afraid of it. That's why you quit.

SOL: Max, I got burnt.

MAX: C'mon, Sol.

SOL: It caused my stroke.

MAX: That's bullshit. It's mathematics, numbers, ideas. Mathematicians are supposed to be out on the edge. You taught me that!

SOL: Max, there's more than math! It's death, Max!

MAX: You can't tell me what it is. You've retreated to your Go and your books and your goldfish, but you're not satisfied.
SOL: Max go home. Get out of my house.
MAX: I'm going to know what it is, I'm going to see it, I'm going to understand it.

4.10 The Name of God

1:07
Synagogue. A rabbi tells Max why the number is important for them.

MAX [incredulous]: You're telling me that the number in my head is the true name of God!?
RABBI: It's more than God—it's everything. It's math and science and nature—the Universe. I saw the Universe's DNA.
MAX: It's just a number. I'm sure you've written down every two hundred sixteen number. You've translated all of them. You've intoned them all. Haven't you? But what's it gotten you?[16] The number is nothing! It's the meaning, the syntax. It's what's between the numbers. If you have not understood it, it's not for you. I've got it, I've got it and I understand it, I'm going to see it! Rabbi—I was chosen.

The 216-digit number being the name of God refers to *Schemhamphoras*, or the *Divided Name of God*. It is hidden in the book of Exodus, chapter 14, verses 19, 20, and 21. Each verse is composed of seventy-two letters (in the original Hebrew). Writing these three verses one above the other, the first from right to left, the second from left to right, and the third from right to left, one gets seventy-two columns of three-letter names of God.

So, contrary to what is said in the movie, this 216-letter name is actually well known. Also, this 216-letter name does not translate into 216 decimals. However, there seems to be (at least) one real secret associated with this 216 letter name; nobody knows how it should be pronounced, and this is apparently essential if you want to tap into its power.

1:11
Sol has died after a second stroke. In Sol's apartment, Max finds a piece of paper on which is written the 216 digit number. Max slides it into his pocket and notices that the pieces on the Go board are arranged in a giant spiral.

4.11 Living Happily Ever After

1:15
Max burns the sheet of paper with the number, then drills into the "math

---

[16] Actually, no mathematician would ever say this, since it is clearly impossible to list all 216-digit numbers, even for a religious fanatic.

section" of his head with a power drill.[17] Afterward, Max can no longer perform the lightning calculations with the little girl Jenna, but he is not bothered. For the first time, we see Max smile. End of a great story.

The ending is summed up perfectly in the director's commentary:

1:16
I always wanted to end the film with little Jenna saying: What's the answer? And having Max not able to give the answer because for the first time he doesn't have the answer. And, in many ways I think that's what π is about. It's about the questions, it's about the chaos, it's about the search for order. That's where Max finds beauty, he finds it now, he finds it in the world around him.

---

[17] "Math section" is the term used in the shooting script.

Chapter 5
Nitpicking in Mathmagic Land

In *Donald in Mathmagic Land* (1959), we accompany Donald Duck (played by himself) as the True Spirit of Adventure introduces him to the wonders of mathematics. This gem of a movie is to mathematics what Disney's *Fantasia* is to classical music. *Donald in Mathmagic Land* is half documentary, half duck action movie.

For decades, *Mathmagic Land* has been an invaluable asset for teachers seeking to inspire their students. However, much of Donald's math is unclear, and some is incorrect. Our intention in this chapter is to sort out what is really going on. We hope it will prove a helpful resource for teachers, and parents, with budding mathematicians to inspire.

Most of the mathematics presented is similar to that in other introductions to the beautiful side of mathematics: the golden section (or golden ratio), the Pythagorean theory of music, conic sections, infinity, and so on. One notable exception is the *diamond rule*, a mathematical rule of thumb for playing three-cushion billiards.

In the following, we'll spend less time on the obvious parts of the movie, with which most readers of this book are probably familiar. Instead, we'll focus upon aspects that are easily overlooked, and we'll do some nitpicking of mathematical claims that are not quite right (or just plain wrong). In particular, we'll draw attention to some of the popular myths about the golden section, and we'll try to make better sense of the diamond rule than is done in the movie.

5.1 Tiny Nitpick: What Is Pi?

Mathmagic Land is populated by plants with number-shaped branches and leaves, trees with square roots (!), and other intriguing mathematical creatures. Particularly memorable are the pencil and pi creatures. The pencil creature has a pencil head and a right-triangled body. It writes random digits on the ground and challenges Donald to a game of tic-tac-toe (which of course the pencil creature wins).

The pi creature recites "π is equal to 3.14159265389747 etcetera, etcetera, etcetera." Surprisingly, he goofs the last two digits: in fact, the decimal expansion of π begins 3.141592653589793... However, the pi creature can at least take comfort in the fact that most other movies featuring a chunk of the decimal expansion of π do even worse; see chapter 18 ("Money-Back Bloopers").

5.2 Historical Nitpick: Pythagoras and the Pythagoreans

To convince Donald that mathematics is "not just for eggheads," the Spirit whisks Donald off to ancient Greece to meet "Pythagoras, the master egghead of them all." The background for the first scene is the painting shown in figure 5.1.

Fig. 5.1 The Pythagoreans: the theorem, some music, and their pentagram logo.

Time for some math spotting! The pentagram at the top was the logo of the Pythagoreans. Underneath, we see what is clearly supposed to be the name "Pythagoras" written in Greek, but there are mistakes: it should be spelled Πυθαγόρας. Then comes the famous diagram that accompanies the first proof of Pythagoras's theorem in Euclid's *Elements*.

This raises an important historical point: while watching the charming Pythagoras, one should keep in mind the uncertainty as to which of "his" discoveries should rather be attributed anonymously to the Pythagorean cult, and which have nothing to do with the Pythagoreans whatsoever.[1] As a telling example, Pythagoras's theorem probably belongs in the latter category![2]

5.3 Small Nitpick: Pythagorean Music

In the painting we also see a number of musical instruments and, in the scenes that follow, the Pythagorean contribution to music theory is sketched. Donald is very impressed that mathematics can be found in music and we witness a musical jam session, with Donald on bongos and Pythagoras on lyre. Then Donald is initiated into the Pythagorean society: Pythagoras stamps a regular pentagram onto Donald's palm.

All great fun! However, this fun also disguises some clumsiness and confusion:

SPIRIT: Pythagoras discovered the octave had a ratio of two to one. With simple fractions he got this [a chord is played upon the lyre]. *And from this harmony in numbers developed the musical scale of today.*

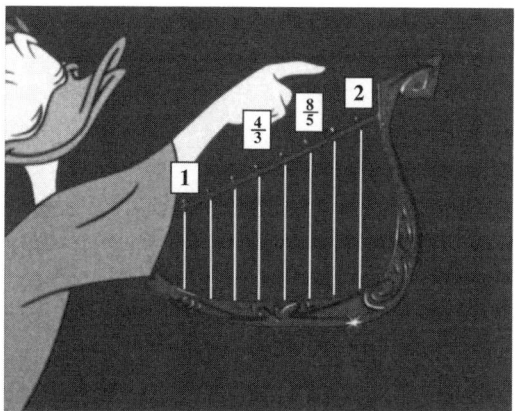

Fig. 5.2 Donald demonstrating the Pythagorean theory of music.

In figure 5.2 we've highlighted the numbers, indicating the string ratios, to which the Spirit refers; the shortest string is half the longest, creating the octave interval. It does not really make sense to say Pythagoras "discovered" that the octave has this ratio, since the octave is simply defined in this

---

[1] For a very thorough study, see Walter Burkert, *Lore and Science in Ancient Pythagoreanism* (Harvard University Press, Cambridge, 1972).

[2] It is clear that even the ancient Babylonians knew of Pythagoras's theorem, long before Pythagoras. And the earliest known proof of the theorem appears in Euclid's *Elements*, dating to about 200 years after Pythagoras's death.

manner; it is (perhaps) more accurate to say that Pythagoras discovered that the ratio of two to one sounds harmonious.[3]

As well, the Pythagorean system of music was not nearly as simple or as obvious as the Spirit suggests. In fact, there are many plausible choices for the system of string lengths, and it was later mathematicians and musicians who placed greater emphasis on the use of simple fractions, an approach that is referred to as *just intonation*. In particular, the Pythagorean system did not use the ratio 8:5, but rather the much less obvious choice of 128:81.[4]

5.4 Medium Nitpick: The Golden Section

The Spirit shows Donald how the golden section arises from the pentagram on Donald's palm;[5] see figure 5.3. Then the Spirit dazzles Donald with some of the most famous of the amazing mathematical properties of the golden section and its omnipresence in nature. Sadly, the Spirit also repeats many of the common but silly myths about the golden section, having to do with the alleged use of the golden section as a major mathematical "law of beauty" in art, in particular in painting and architecture. Almost everything shown and said for two minutes from the nine minute mark is nonsense:[6]

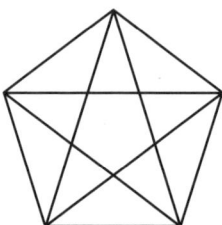

Fig. 5.3 The golden section equals the ratio of a diagonal of the pentagon to a side.

SPIRIT: To the Greeks, the golden section represented a mathematical law of beauty. We find it in their classical architecture. The Parthenon one of the most famous Greek buildings contains many golden rectangles.

In the movie, and whenever such claims are made, the only evidence provided is a golden rectangle or two superimposed onto the object in question. Figure 5.4 is a very famous example involving the Parthenon. At first glance

---

[3] Again, it is not clear whether Pythagoras should really be credited with this discovery.

[4] For a lovely exposition of the history and mathematics of musical scales, see Trudy Garland and Charity Kahn, *Math and Music* (Dale Seymour, Palo Alto, CA, 1995).

[5] For a brief introduction to the golden section (or golden ratio), see chapter 4 ("The Annotated π Files").

[6] For some thorough debunking, see Mario Livio's book, referenced in footnote 10 of chapter 4, and George Markowsky, Misconceptions about the golden ratio, *College Mathematics Journal* 23 (1992), 2–19.

this may appear quite convincing, but on closer inspection it is clear that other rectangles could have been equally well chosen. For example, there is no particular reason why the bottom of the superimposed rectangle should end where it does, rather than at the bottom of the columns or the bottom of the stairs.

Fig. 5.4 The Parthenon exhibiting perfect proportions.

SPIRIT: The same golden proportions are also used in their sculpture. In the centuries that followed, the golden section dominated the idea of beauty in architecture throughout the western world. The cathedral of Notre Dame is an outstanding example.

Although the golden section arose in the geometry of the ancient Greeks, there is no concrete evidence of its application in art or architecture prior to Luca Pacioli's linking of the golden section to aesthetics in the early sixteenth century. Even then, the earliest definite application of the golden section in art was not until the ninteenth century.

SPIRIT: The Renaissance painters knew this secret well.

Here we are shown Leonardo da Vinci's *Mona Lisa*, cluttered with super-imposed rectangles. Leonardo was certainly aware of the golden section and even acted as illustrator for Luca Pacioli. However, it is highly questionable whether Leonardo used it in his own creations. If so, he never bothered to mention it in any of his writings.

SPIRIT: Today, the golden rectangle is very much a part of our modern world.

At this point we are shown the facade of the United Nations headquarters, divided up by three supposedly golden rectangles. However, measurements of these rectangles show that they are not close to golden.

SPIRIT: Modern painters have rediscovered the magic of these proportions.

A few have, such as Salvadore Dali: most haven't. Our discussions with architects and artists suggest that they tend to care about the golden section to the exact extent of feeling confused and guilty. And we strongly suspect that those few who now employ the golden section do so with no clear aesthetic purpose and merely from the mistaken belief that they're following a classical tradition. That is, the supposed aesthetic beauty of the golden section has become a self-perpetuating myth.

SPIRIT: Indeed, this ideal proportion is to be found in nature itself.

Here, we are shown a beautifully proportioned ballerina, with several golden rectangles chopping up her body. Again, it is no more than wishful thinking that golden section proportions are the key to beauty in the human body.

By the end of this golden section marathon, Donald is inspired to demonstrate that his body, too, has perfect proportions. He succeeds, by jamming himself into a regular pentagon (see figure 5.5), and the Spirit laughs at Donald.

Fig. 5.5 Donald exhibiting perfect proportions.

The Spirit might have kept his laughter in check if he had listened to the speech on numerology in the movie π (see section 4.7). Then the Spirit could have realized that squeezing the Parthenon into a golden rectangle is no different from Donald's contortionist exercise. A step further, and the Spirit could have realized that with sufficient work and by turning a blind eye or two, one can find the golden section, or any number desired, pretty much anywhere.

This interesting but very misleading section is followed by an impressive slideshow of natural shapes that, for the most part, really *are* based upon the golden section: pentagonal flower heads, branching patterns, seed heads, pinecones, and so on. Alas, the Spirit then repeats the enticing but groundless myth that the spirals of nautilus shells are related to the golden ratio.

5.5 Large Nitpick: Three-Cushion Billiards

The Spirit points out that mathematics is found in many games. First, chess: Donald turns into Lewis Carroll's Alice and makes a narrow escape. Then, baseball, football, basketball, and hopscotch. All of these examples are rather weak. For example, hopscotch is apparently mathematical because of its "multiple squares." But, finally, we have:

SPIRIT: A mathematical game played on a field of two perfect squares, using three perfect spheres, and a lot of diamonds. In other words, billiards.

It is clear that Donald is being guided by an American Spirit, and that by "billiards" he means three-cushion billiards, rather than the more common English billiards. Both games use three balls, but a three-cushion billiard table has no pockets. In three-cushion billiards, the aim is to make the cue ball hit the other two balls, with the cue ball contacting at least three rails (or cushions) before it hits the second target ball.

The diamonds divide the long side of the table into eight equal intervals and the short side into four. We'll explain how an expert player uses the diamonds to plan a typical shot, such as that shown in figure 5.6. To simplify the following discussion, we'll just consider two balls, with the white ball the cue ball and the black ball the target ball.[7]

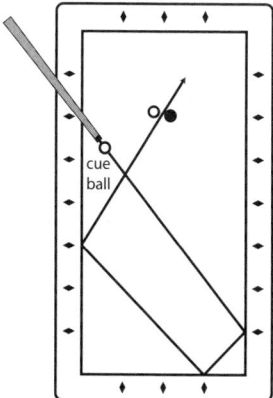

Fig. 5.6 Making a three-cushion shot.

Figure 5.7 shows one of the billiard shots discussed in the movie. Here, A marks a point on the left rail, the imaginary point where the cue ball began. Then B and C mark the points at which the cue ball bounces off the two long rails.

---

[7] For an excellent discussion of the diamond system, see G. L. Cohen, Three cushion billiards: notes on the diamond system, *Sports Engineering* 5 (2002), 43–51.

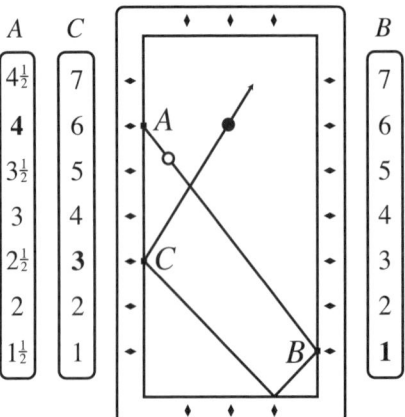

Fig. 5.7 The diamond rule $A - B = C$: here, $A = 4$, $B = 1$, and $C = 3$.

The diamonds on the left side of the table are numbered in two ways, first for point A and then for point C, as indicated in figure 5.7. Similarly, the diamonds on the right side of the table are labeled for point B. So, in this example, $A = 4$, $B = 1$, and $C = 3$. Then, the *diamond rule* is $A - B = C$. In our example, this amounts to $4 - 1 = 3$.

It is important to realize that the mathematics of real billiard balls is very complicated, and that the diamond rule is no more than a rule of thumb. In particular, sidespin of the cue ball means that the ball will normally bounce off a cushion at a greater angle than the angle of approach.

So, the diamond "rule" $A - B = C$ is only a guide. However, the rule is sufficiently accurate for planning many shots on a real billiard table. As in the movie, we'll make things easier by pretending that $A - B = C$ holds exactly, and we'll ignore the subtleties of its real-life application.

We now describe how a player can apply the diamond system to plan a shot such as that shown in figure 5.7. In the simplest situation, both the white and black balls are against the left rail, coinciding with the points A and C respectively. Then, since $A = 4$ and $C = 3$, we calculate $B = A - C = 1$. This indicates that, if we aim the white ball at the first diamond on the bottom right, the white ball should exactly follow the path indicated in figure 5.7 and hit the black ball at C.

In a more complicated situation, the white ball could be somewhere in the middle of the table with the black ball against the left rail. To find the values of A and B, the player swivels the cue above the white ball, and keeps subtracting the right (B) end of the cue from the left (A) end, until the resulting subtraction gives the value C; see figure 5.8.

In our example, $C = 2$ and the player's guesses are

$$A_1 - B_1 = 4\frac{1}{2} - 1 = 3\frac{1}{2},$$

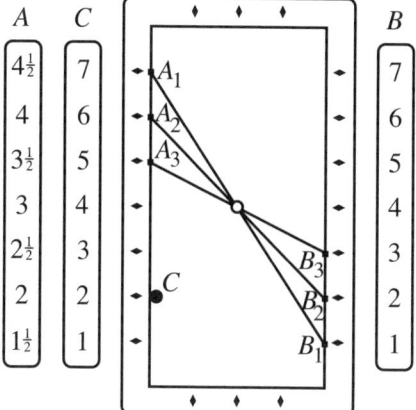

Fig. 5.8 Swivel the cue through the white ball until $A - B = C$.

$$A_2 - B_2 = 4 - 2 = 2\,,$$

$$A_3 - B_3 = 3\frac{1}{2} - 3 = \frac{1}{2}\,.$$

The second subtraction gives the actual value of C, and so the diamond rule instructs us to aim at the second diamond at the bottom right.

Finally, the trickiest situation occurs when the black ball is also in the middle of the table; see figure 5.9. Here, there is no methodical system using the diamonds to determine the correct values for A, B, and C. However, an experienced player, with an intuitive feel for the way balls bounce off the rails, will have a good idea of how the angle at C depends on the angle at A.

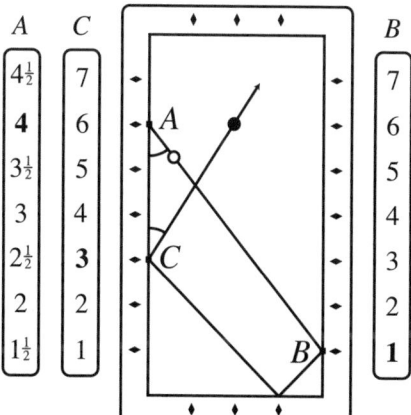

Fig. 5.9 The most complicated situation.

Here is how an experienced player can use the swivel technique together with intuition for the angle relationship to determine the correct values. Every swivel position corresponds to an A and a B, together with an angle at A. The player then calculates C via the diamond rule and guesses the angle at C based on the angle at A. Once the position of C and the angle at C are in line with the black ball, he knows that the values are correct and can make his shot.

We can see that the diamond system is a beautiful application of some simple (and some complicated) mathematics, but the Spirit's exposition would hopelessly confuse anybody. Here is the relevant part of the Spirit's explanation, which refers to the complicated situation illustrated in figure 5.9.

SPIRIT: He uses the diamond markings on the rail as a mathematical guide. First he figures the natural angle [at C] *for hitting the object balls* [the black ball]. *And then he finds that his cue ball* [the white ball] *must bounce off the no. 3 diamond. Next he gets ready for the shot, and he needs a number for his cue position* [A]. *This calls for a different set of numbers* [different labeling]. *You see, the cue position* [A] *is 4. Now, a simple subtraction $4 - 3 = 1$. So, if he shoots for the first diamond* [B], *he should make it ...*

In effect, the Spirit is suggesting that it is easy to determine the points C and A in figure 5.9 without using the swivel technique, and that the player only needs to perform a simple subtraction to determine B and so guarantee a successful shot.

Of course this makes no sense, because if the white ball is in the middle of the table, then its position together with that of A determines the shot, and so B and C would already be determined, without using the diamond rule at all! What is even funnier is that when Donald tries to use the diamond system, he is in effect applying the swivel technique and gets scolded by the Spirit for doing so.

There are a number of other technical issues that the Spirit ignores. First, as mentioned above, the diamond rule is really just a rule of thumb and is not mathematically infallible as suggested by the Spirit. Second, for the diamond rule to work well, a player is supposed to provide the cue ball with a certain amount of sidespin so that the rebound angles are predictably larger than the approach angles (as is the case in all our diagrams so far). Third, the diamond rule as described in the movie and by us is only part of a more comprehensive diamond system, which also covers cases in which A has to be chosen somewhere along the top rail. Fourth, according to many modern books on billiards, the diamond system works best if you aim directly at the diamonds and not at the rail positions adjacent to the diamonds. Finally, taking into account that the balls have nonzero radius would require slight alterations to all our diagrams.

It is unavoidable that rules for real billiards can only be rules of thumb. On the other hand, it is insightful and fun to figure out what the diamond system would be in an ideal game of *mathematical billiards*. If we assume that the rebound and approach angles of the cue ball are always equal then precise rules are indeed possible. And if we further assume that the cue ball has zero radius, the rules take a very simple form. Using the same labelings as before, we get a *precise* diamond rule for mathematical billiards:

$$A - B = 1 + \frac{C}{2}.$$

For example, in figure 5.10 we have $A = 4$, $B = 1$ and $C = 4$. And, indeed $4 - 1 = 3 = 1 + \frac{4}{2}$.

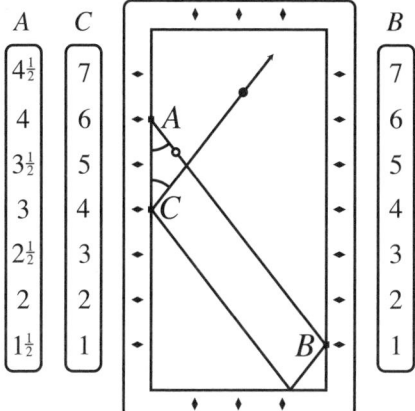

Fig. 5.10 The angles at A and C are equal, and the diamond rule becomes $A - B = 1 + \frac{C}{2}$.

In mathematical billiards, we can use exactly the same strategies as those outlined above to plan our shots. Even the situation in which both balls are off the rails is easy. This is because, as is readily checked, the angles at C and A are always equal.

As a final simplification, we can eliminate the peculiar numbering for A, and use the "natural" numbering of the diamonds for all of A, B, and C; see figure 5.11. Now the diamond rule for mathematical billiards takes the form

$$A - 2B = C.$$

It is easy to see why this last rule should be correct. Since the incoming and outgoing angles are always equal, we can unfold the path of the white ball into a straight line. This is shown in the left diagram of figure 5.12, with the new target point C indicated relative to the reflected numbering. It is also clear that the highlighted right-angled triangles are identical. This means that the lengths of the two vertical sides of these triangles are equal, and so $A - B = C + B$. That is, $A - 2B = C$. Q.E.D.

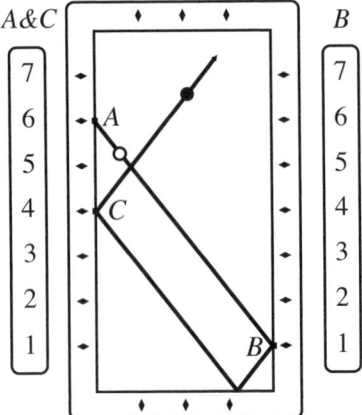

Fig. 5.11 After simplifying the labeling for A, the diamond rule becomes $A - 2B = C$.

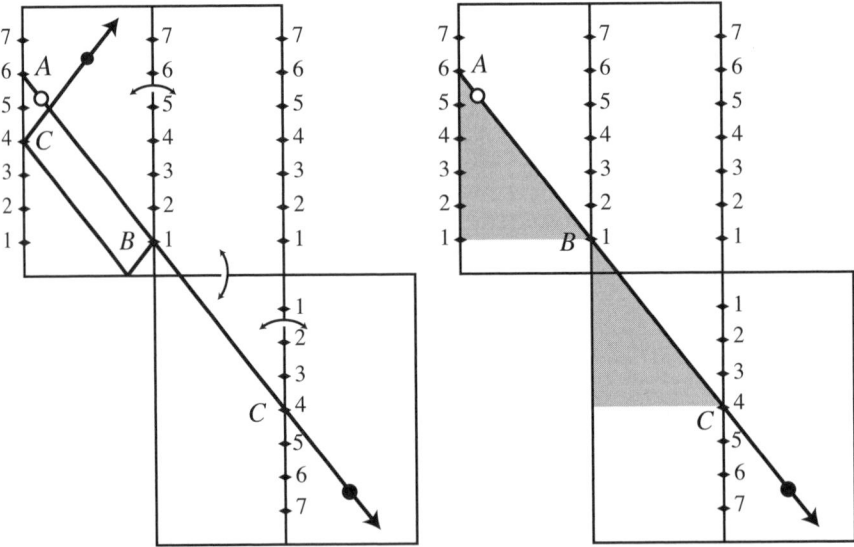

Fig. 5.12 Deriving the mathematical diamond rule $A - 2B = C$ by unfolding the path.

5.6 NOTpick

Our choice of focus in this chapter has the consequence that we ended up saying a number of not-so-nice things about *Mathmagic Land*. This may leave the impression that we don't like the movie, or that it is not worth watching. Definitely, neither is the case. *Donald in Mathmagic Land* is an engaging and inspirational movie about mathematics, a must-see for every budding mathemagician. As the Spirit elegantly summarizes at the end of the movie:

In the words of Galileo: mathematics is the alphabet with which God has written the Universe.

Donald in Mathmagic Land does a very fine job of illustrating Galileo's words.

5.7 Notes

Consultants: Milt Banta, Bill Berg, and Heinz Haber are credited for their contribution to the development of the story of Mathmagic Land. Most likely it was the astrophysicist Dr. Heinz Haber (1913–1990) who was responsible for the math. As a scientist, he became famous for being one of the founders of a new field of research: space medicine. He was also one of the main science popularizers of his time and was involved in several of Disney's other educational productions.

Mathematics and Billiards: Other movies in which pool or billiards is played "mathematically" include *Lambada* (1990) and *Little Man Tate* (1991). The billiards scene in *Lambada* makes an amusing supplement to any screening of *Mathmagic Land*. Most notable is the scene in which the hero teacher Blade plans his three-cushion shot using the diamonds and a protractor (figure 5.13)!

Fig. 5.13 Blade, the cool math teacher in *Lambada*, using his trusty protractor.

Comic Book: Every true fan of *Donald in Mathmagic Land* should also try to get hold of a copy of the comic book of the same title which appeared in 1959 (Dell comic no. 1051). However, be warned, although it cost only 10 cents when it was published, our copy cost us $30 on eBay.

The comic book covers both more and less mathematics than the movie. Notably, the pi creature gets π correct: "Pi is the Greek letter used to designate the relation of the circumference of a circle to its diameter! Pi is three point one four one five nine two six five."

Donald visits some Stone Age people, as well as ancient Greece, and finds out the hard way why it is useful to be able to count. He also learns about the decimal system and some of the other ways that the ancient people used to note down numbers. Billiards receives a mention, but the diamond system is only hinted at.

In the central story of the comic, Donald meets the mathemagician Nimble Numbo, who shows Donald how to beat Uncle Scrooge and get out of his debt. Here is Donald putting the plan into action (we've streamlined the text a little bit):

The balance I still owe you is about fifteen dollars, by now, I expect. So, I'll give you my whole house to pay it off! You'll have to pay me a little something extra to make up the difference, naturally [Donald gets out a chessboard]. *At least a few pennies, Uncle Scrooge! Here! Here's your chessboard!*

"I'll settle if you'll just put a penny on the first square and two pennies on the second square and four pennies on the third. Just double the number of pennies each time until you've given me some for every one of these sixty-four squares.

Of course, the result is $2^{64} - 1$ cents, the amount of which comes as a very unpleasant surprise to Uncle Scrooge.

More Math in Disney Movies: Mickey Mouse gets tortured with trigonometry in the *Prince and the Pauper* (1990). Also, in *Dumbo* (1941), Dumbo manages to blow a cubical bubble, a feat that is actually possible, and with which mathematicians love to amaze their students.

Chapter 6
Escape from the Cube

In *Cube* (1997), some very unlucky people wake up in a deadly labyrinth, the Cube, and have to find a way out. Leaven, a math student, does most of the mathematical deciphering. She is assisted by Worth, one of the engineers involved in building the Cube, and Kazan, who is autistic and very good at factoring numbers. The other characters are Alderson, an early victim of the Cube; Quentin, a cop; Rennes, an escape artist; and Holloway, a doctor.

6.1 The Cube in *Cube*

The mathematics behind the Cube really works. In fact, the Cube was designed by David W. Pravica, a professor of mathematics at East Carolina University.[1] Pravica was also consulted during the making of the movie, but some inconsistencies were introduced during the shooting and editing.

The Cube is a 3D counterpart of the 2D diagram in figure 6.1. It consists of an outer cubical shell, the sarcophagus, and an inner cubical shell. The sidelengths of the outer shell are 434 feet. The inner shell is subdivided into $26 \times 26 \times 26 = 17,576$ smaller cubical spaces, each with sidelength 15.5 feet.

The distance between the outer and inner shells of the Cube is also 15.5 feet, and usually this space is empty. At any time, a cubical space is either empty or is occupied by a cubical room. Some of the rooms are booby trapped. Periodically, some of the rooms slide to new locations. It is not clear in the movie whether the rooms are meant to move simultaneously: we discuss this in section 6.5.

Each cubical room has six identical square doors, one in the center of each wall, through which the Cube's captives can move to an adjacent room. At regular time intervals exactly one of the rooms, the bridge, moves out of the

---

[1] For some more interesting reading on *Cube*, and the Cube, see Michele Emmer and Mirella Manaresi, *Mathematics, Art, Technology, and Cinema* (Springer, New York, 2003). In particular, note the article by David Pravica and Heather Ries on the mathematics of the Cube, and the interview with Vincenzo Natali, the director of *Cube*.

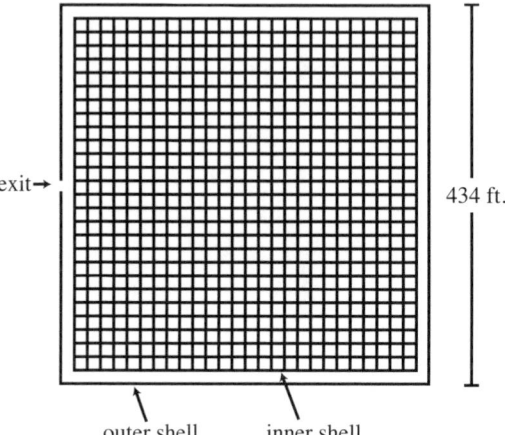

Fig. 6.1 The two-dimensional counterpart of the Cube.

inner shell and in front of the only exit in the outer shell. So, to escape the Cube, a captive has to get to the bridge and then wait until the bridge moves in front of the exit; or they must find a room that will be adjacent to the bridge when the bridge is in front of the exit.

Rennes dies early in the movie (best not to ask how). After his death, the other captives conclude that it would be helpful to have a safe method of identifying booby-trapped rooms. It is possible for them to do this because the builders of the Cube encoded information about the geometry of the Cube, and whether or not a room is safe, into numerical labels in each of the rooms. Furthermore, each captive has a special ability or possesses critical information about the Cube. To begin with, the characters don't suspect any of this.

In the following, we'll have a closer look at how the characters figure out (the hard way) what is going on.

6.2 First Insight: The Power of Primes

0:11

The characters have noticed the number labels 566, 472, and 737 in one room and the number labels 476, 804, and 539 in an adjacent room (figure 6.2).

QUENTIN: What is it, serial numbers?
HOLLOWAY: Room numbers, they're different in each room.
WORTH: Oh great, so there's only five hundred and sixty-six million, four hundred thousand rooms in this thing.
HOLLOWAY: There'd better not be! We have about three days without food and water before we're too weak to move.

Fig. 6.2 One of the number triples.

0:18

QUENTIN: Leaven, what do you do in school? Math?
HOLLOWAY: What can they mean?

Leaven puts on her glasses and begins to study the numbers.

LEAVEN: 149?

Leaven opens a new door. The numbers we see are 645, 372, and (later) 649.

LEAVEN: Prime numbers. I can't believe I didn't see it before.
QUENTIN: See what?
LEAVEN: It seems like if any of these numbers are prime, then the room is trapped. Okay, 645—645, that's not prime. 372—no. 649—wait, 11 times 59, it's not prime either. So, that room is safe.

It is unintentionally hilarious that Leaven hesitates before declaring that 645 and 372 are not prime: since the last digits are 5 and 2, it's not as if involved calculations are required. Then, after struggling over the easy ones, Leaven figures out that $649 = 11 \times 59$ almost immediately.

QUENTIN: Wait, wait, wait. How can you make that assumption based on one prime number trap?
LEAVEN: I'm not. The incinerator thing was prime: 083. The molecular-chemical thingy had 137, the acid room had 149.
HOLLOWAY: You remembered all that in your head?
LEAVEN: I have a facility for it.

6.3 How to Avoid Prime Numbers

In the rest of the movie, we repeatedly see Leaven, the math expert, trying to figure out whether or not a three-digit number is prime. So, if you woke up tomorrow and found yourself in the Cube, how would you go about this task? As a public service, we now provide a prime number survival guide.

A three-digit number has the form abc: for example, the "molecular-chemical thingy" had label 137, for which $a = 1$, $b = 3$, and $c = 7$.[2] Now, if abc is not prime then it has at least one prime factor no greater than its square root. (Otherwise, there would have to be two larger factors, and their product would be too big.) And, since abc is at most 999, which has square root about 31.6, we have to check for divisibility by the primes up to at most 31.

First, we quickly check whether abc is one of the primes up to 31:

$$2, 3, 5, 7, 11, 13, 17, 19, 23, 29, 31.$$

Next, we check whether abc is an even number, that is, whether it is divisible by 2: that caused Leaven some difficulty, but of course it's not actually that tough. If abc is an even number then it is not prime, and we are done.

Then we check whether abc is divisible by 3, then for divisibility by 5, then by 7, and so on, up to the largest prime below $\sqrt{abc}$. As soon as we find a factor we stop, and if we fail to find a factor then abc must be prime.

We now know what we have to do. To actually get through the work quickly, before the Cube does something evil, we just have to employ some well known, and some less well known, divisibility tricks:

2: The number abc is divisible by 2 if its last digit c is 0, 2, 4, 6, or 8. (Leaven, please take note).

3: The number abc is divisible by 3 if the sum $a + b + c$ is divisible by 3. For example, 543 is divisible by 3 since $5 + 4 + 3 = 12$, and 12 is divisible by 3.

5: The number abc is divisible by 5 if its last digit c is 0 or 5. (Another one for Leaven to practice.)

7: The number abc is divisible by 7 if $2a + 3b + c$ is divisible by 7. For example, 364 is divisible by 7 since $(2 \times 3) + (3 \times 6) + 4 = 28$ is divisible by 7.

11: The number abc is divisible by 11 if $a - b + c$ is divisible by 11. For example, 649 is divisible by 11 since $6 - 4 + 9 = 11$ is. (Leaven seems to have practiced this trick.)

In fact, there is a divisibility trick for any prime number. Of course, the vast majority of the numbers up to 999 are divisible by 2, 3, or 5. So just the easy divisibility tricks for these three smallest primes should protect us from a lot of Cube traps.

As some final practice before returning to the Cube, let's determine whether 137 is prime. The square root of 137 is a bit greater than 11. Using our divisibility tricks, we quickly convince ourselves that 137 is not divisible by 2, 3, 5, 7, or 11, and so we conclude that 137 is a prime number.

---

[2] By "three-digit number" we're including the possibility that the digits in the hundreds and tens places might be 0. For example, the incinerator thing had prime label 083, for which $a = 0$.

For a while, Leaven's prime number test works perfectly. However, at some point Quentin almost gets killed in a room whose numbers are not prime. Before Leaven figures out what is going on, the captives stumble across another critical piece of information.

6.4 Second Insight: The Cube in Coordinates

0:39
Worth has confessed that he designed the outer shell of the labyrinth, and that he is aware that the overall shape is a cube. However, he doesn't know anything about the structure of the inner shell.

LEAVEN: What are the dimensions of the outer shell?
WORTH: 434 feet square.

Leaven paces out the room they're in, to determine its dimensions.

LEAVEN: 14 by 14 by 14.
WORTH: The inner cube cannot be flush by the shell. There is a space.
LEAVEN: One cube?
WORTH: I don't know. It makes sense.
LEAVEN: Well, the biggest the cube then can be is—26 rooms high, 26 rooms across, so—17,576 rooms.

Later, Leaven figures out the importance of the cube being 26 rooms across. Notice that dividing 434 by 14 gives 31, suggesting that $31 \times 31 \times 31$ rooms fit into the outer shell, with $29 \times 29 \times 29$ spaces in the inner shell. However, this does not take into account the thickness of the walls. Assuming the Cube is 28 spaces across and working backward, this suggests that the walls are about 1.5 feet thick, which seems reasonable.

HOLLOWAY: 17,576 rooms? Oh God, that makes me queasy.
LEAVEN: Descartes.

Leaven opens a new door and puts on her glasses.

LEAVEN: Leaven, you are a genius!
QUENTIN: What?

We see three numbers: 517, 478, and 565.

LEAVEN: Cartesian coordinates, of course, coded Cartesian coordinates. They're used in geometry to plot points on a three-dimensional graph.
QUENTIN: In English. Slower.
LEAVEN: Bonjour! These numbers are markers, a grid reference, like latitude and longitude on a map. The numbers tell us where we are inside the cube.
QUENTIN: Then where are we?

Leaven has just figured out that, as well as indicating traps, the numbers in each room also encode its coordinates. The x-coordinate is the sum of the digits in the first number, the y-coordinate is the sum of the digits in the second number, and the z-coordinate is the sum of the digits in the third number. For example, the coordinates of the room with identification numbers 517, 478, and 565 are

$$(5 + 1 + 7, 4 + 7 + 8, 5 + 6 + 5) = (13, 19, 16).$$

LEAVEN: It works! The x-coordinate is 19.

Here, Leaven scribbles 928 on a piece of metal. This adds to $9 + 2 + 8 = 19$, but she is presumably not working on the triple 517 478 565 we were just shown.

LEAVEN: y is ...

She scribbles 856, giving $y = 8 + 5 + 6 = 19$.

LEAVEN: 26 rooms. So [because $26 - 19 = 7$], *that places us—seven rooms from the edge.*

0:42
Leaven is puzzling over a new set of coordinates.

QUENTIN: What's the matter?
LEAVEN: These coordinates: (14, 27, 14).

Notice that the only way to get a y-coordinate of 27 is if the second identification number is 999. So 27 is definitely the largest coordinate that we will ever come across. However, ...

QUENTIN: What about them?
LEAVEN: Well, they don't make sense. Assuming the cube is 26 rooms across, there can't be a coordinate larger than 26. If this were right, then we would be outside the cube.

This observation will prove to be very important.

6.5 Third Insight: Permutations

1:04
The captives try to return to the room containing Rennes's body. Rennes had been killed in an adjacent room, which has since disappeared.

WORTH: Wasn't Rennes killed in that room?

Worth opens the door to where Rennes was killed, but there is nothing. All we see is black. It's the outer shell.

WORTH: How come there's nothing out there?

WORTH: Hey! Listen to what I'm saying. There was a room there before. We haven't been moving in circles, the rooms have!

LEAVEN: Of course—It's the only logical explanation. I'm such an idiot.

WORTH: What are you on to, Leaven?

LEAVEN: Give me a minute. The numbers are markers, points on a map, right?

WORTH: Right.

LEAVEN: And how do you map a point that keeps moving?

WORTH: Permutations.

QUENTIN: Permu—what?

LEAVEN: Permutations. A list of all the coordinates that the room passes through. Like a map that tells you where the room starts, how many times it moves, and where it moves to.

It turns out that the three coordinates don't give the present location of a room, only its starting location. However, Leaven has figured out that a room's identification numbers also encode the room's movement through the Cube. We'll explain.

To determine the subsequent locations of a room, we first calculate, for each identification number abc of the room, the following triple of numbers:

$$a - b, \quad b - c, \quad c - a.$$

Consider the room labeled 665 972 545, where our characters will shortly find themselves. We first calculate

$$665 \ \rightarrow \ 6 - 6 = 0, \ 6 - 5 = 1, \ 5 - 6 = -1.$$

So the first triple of numbers is $0, 1, -1$. Then,

$$972 \ \rightarrow \ 9 - 7 = 2, \ 7 - 2 = 5, \ 2 - 9 = -7$$

and the second triple of numbers is $2, 5, -7$. Finally,

$$545 \ \rightarrow \ 5 - 4 = 1, \ 4 - 5 = -1, \ 5 - 5 = 0.$$

The third triple of numbers is $1, -1, 0$.

To see how these triples govern the movement of the room, we'll follow the room on its journey. To begin, we add the digits of each identification number, and find our room has starting location

$$(6 + 6 + 5, 9 + 7 + 2, 5 + 4 + 5) = (17, 18, 14).$$

Now, for the first move, add the first number of the first triple to the x-component, giving

$$(17 + 0, 18, 14) = (17, 18, 14).$$

So we haven't gone anywhere. However, for the second move, add the first number of the second triple to the y-component, giving

$$(17, 18 + 2, 14) = (17, 20, 14).$$

For the third move, add the first number of the third triple to the z-component, giving

$$(17, 20, 14 + 1) = (17, 20, 15).$$

Now, repeat this procedure with the second numbers of each triple, and finally with the third numbers of each triple. So the path continues

$$\rightarrow (18, 20, 15) \rightarrow (18, 25, 15) \rightarrow (18, 25, 14)$$
$$\rightarrow (17, 25, 14) \rightarrow (17, 18, 14) \rightarrow (17, 18, 14).$$

In all, our room has moved nine times, though the first and last "moves," corresponding to the 0s, actually amount to remaining still. At the end of the nine moves, the room has returned to its starting location, and it then cycles through these moves, over and over.

In fact, each room will have a similar cycle, where it returns to its starting location after every nine moves. This is a direct consequence of the sum of the shifts for a room coordinate being $(a - b) + (b - c) + (c - a) = 0$.

Having figured out the way rooms move about, let's return to see how Leaven is coping:

QUENTIN: *The number tells you all that?*
LEAVEN: *I don't know. See, I've only been looking at one point on the map, which is probably the starting position. All I saw was how the cube looked like before it started to move* [the initial coordinates of the room].
QUENTIN: *Okay, so it's moving. How do we get out?*
LEAVEN: *27. I know where the exit is. You remember that room we passed through before, the one with the coordinate larger than 26?*
WORTH: *What about it?*
LEAVEN: *That coordinate placed the room outside the cube.*
WORTH: *A bridge?*
LEAVEN: *Right, but only in its original position.*
QUENTIN: *What are you talking about?*
LEAVEN: *Look, the room starts off as a bridge. Then it moves its way through the maze, which is where we ran into it. But at some point it must return to its original position.*
WORTH: *So the bridge is only a bridge . . .*
LEAVEN: *For a short period of time. This thing is like a giant combination lock. When the rooms are in their starting positions the lock is open. But when they move out of the alignment the lock closes.*
QUENTIN: *So when does it open?*
WORTH: *For a structure this size—it must take days to complete a full cycle.*

We see Leaven calculating. They are now supposed to be in the room with identification numbers 665, 972, and 545.[3]

LEAVEN: We find its original coordinates by adding the numbers. The permutation is found by subtracting the numbers. That's it. The room moves to 0, 1, and −1 on the x-axis, 2, 5, and −7 on y, and 1, −1, and 0 on z.

Leaven has correctly figured out the cycle of movements that we calculated earlier. However, she is about to go further.

QUENTIN: And what does that mean?
LEAVEN: You suck in math? Okay, I need the room numbers around as a reference point.

Leaven has just figured out that it is possible for her to determine the current position of the room they're in. The trick is to compare the room's cycle of coordinates with those of an adjacent room.

WORTH: 666—897—466.
QUENTIN: 567—898—okay?
LEAVEN: Yes!
QUENTIN: And 545—Did you get that?

We, and Leaven, have already calculated that the room they're in, with identification 665 972 545, has the cycle

$$\text{start} = (17, 18, 14) \to \text{same} \to (17, 20, 14) \to (17, 20, 15) \to (18, 20, 15)$$
$$\to (18, 25, 15)^* \to (18, 25, 14)^\dagger \to (17, 25, 14)^\ddagger \to (17, 18, 14) \to \text{same} = \text{start}.$$

Now, compare this to the cycle of Worth's adjacent room. Calculating as above, we find that this room cycles through the locations

$$\text{start} = (18, 24, 16) \to \text{same} \to (18, 23, 16) \to (18, 23, 14) \to \text{same}$$
$$\to (18, 25, 14)^{*\ddagger} \to \text{same}^{*\ddagger} \to \text{same}^{*\ddagger} \to (18, 24, 14)^\dagger \to (18, 24, 16) = \text{start}.$$

If two rooms are adjacent then their coordinates must coincide in two entries and differ by 1 in the remaining entry. This implies that Leaven and her comrades must be at one of three locations indicated by a superscript, with the possible locations of Worth's room indicated by the corresponding superscripts.

We now have to consider a subtle issue of how the Cube was intended to work. The simplest approach would have been to have all the rooms move in sync, which is possibly what was originally intended.[4] If this were the case, then adjacent rooms would not only have comparable coordinates, as we have already indicated, they would moreover be at the same stage of their cycles.

---

[3] We are not told this in the movie, but it is indicated in David Pravica's and Heather Ries's article, referred to in footnote 1.

[4] See David Pravica and Heather Ries's article referred to in footnote 1.

For Leaven and co., that leaves two possibilities: they are at $(18, 25, 15)^*$, with Worth's room being at $(18, 25, 14)^{*\ddagger}$ (five moves into the cycle); or, they are at $(17, 25, 14)^\ddagger$ with Worth's room at $(18, 25, 14)^{*\ddagger}$ (seven moves into the cycle).

Unfortunately, the Cube in the movie cannot be that simple. If it were, Leaven's room would have crashed into Worth's room at $(18, 25, 14)$, during the sixth move of the cycle. So, either it was intended for the room movements to be out of sync, or this was an error that slipped in during film production.[5]

Whatever the case, at this stage Leaven needs more information. This has been provided by Quentin. The identification numbers for his room are 567, 898, and 545, which gives the coordinate cycle

$$\text{start} = (18, 25, 14)^{*\ddagger}$$
$$\to (17, 25, 14)^\dagger \to (17, 24, 14)^\ddagger \to (17, 24, 15) \to (16, 24, 15)$$
$$\to (16, 25, 15) \to (16, 25, 14)^\ddagger \to (18, 25, 14)^{*\ddagger} \to \text{same}^{*\ddagger} \to \text{same}^{*\ddagger} = \text{start}.$$

Now, if the above room movements are meant to be in sync, a comparison of coordinates indicates that the only remaining possibility is that Leaven's room is at $(17, 25, 14)^\ddagger$, seven moves into the cycle. Alas, this still cannot work: Worth's and Quentin's rooms are indeed just a step away, but they have also collided at $(18, 25, 14)^{*\ddagger}$.

Anyway, the director neglected to inform Leaven about colliding rooms, and so she carries on calculating. However, even if the room movements are not in sync, Leaven has sufficient information to determine their location. From the above coordinate cycles there are only two possibilities: either Leaven and the gang are at coordinates (17, 25, 14), with Quentin and Worth peeking into (18, 25, 14) and (16, 25, 14); or they are at (18, 25, 14), with Quentin's and Worth's rooms at (18, 24, 14) and (17, 25, 14). And these two possibilities are distinguishable, since in the first scenario the three rooms are in a line, and in the second scenario the rooms form an L-shape (figure 6.3).

We cannot actually tell from the movie how the three rooms are positioned. However, eventually Leaven in fact concludes that they're at coordinates (17, 25, 14). To do this, she requests more information:

WORTH: 656—778—462.

This is supposed to identify a third adjacent room, but a calculation of the room's coordinate cycle shows that this is impossible. Another blooper.

LEAVEN: That's enough. x is 17, y is 25, and z is 14. Which means this room makes two more moves before returning to its original position.
WORTH: Do we have time?
LEAVEN: Maybe.

---

[5] When we talked to David Pravica, he suggested that the Cube was *not* intended to have synchronized room movements.

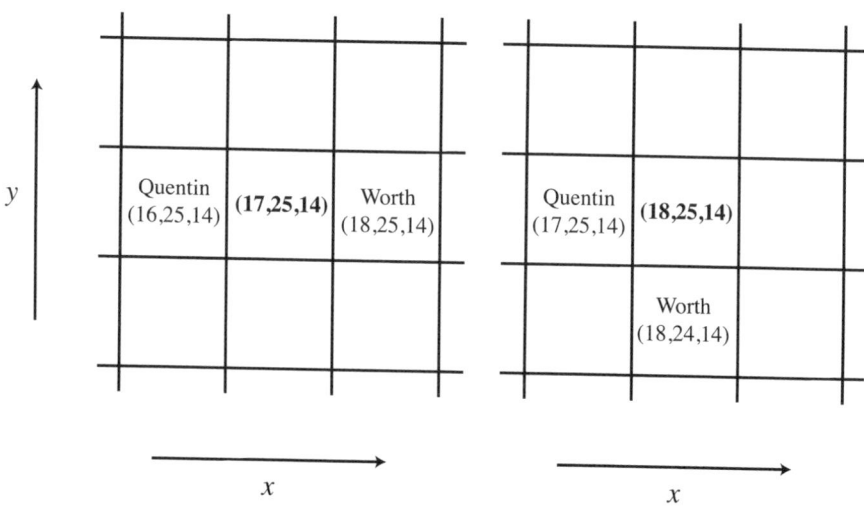

Fig. 6.3 The current position of their room is either $(17, 25, 14)$ or $(18, 25, 14)$.

In fact, their room is just one move away from its starting position, since the last "move" has no effect. True, Leaven might be calculating that the Cube as a whole needs two moves to return to its starting configuration. However, that makes only sense if the rooms move in sync. Indeed, at this stage, Leaven's solution of the puzzle makes sense only under that assumption. In which case, we'll have to deal again with all those crashing rooms.

6.6 Final Insight: Prime Powers

QUENTIN: Then let's go.
WORTH: Can you work out the traps in the system?
QUENTIN: Fuck the traps, let's get to the bridge.
WORTH: Well, you threw out our last boot, you fucking idiot.
LEAVEN: Technically I can identify the traps.
WORTH: Technically?
LEAVEN: First I thought they were identified by prime numbers, but they're not. They're identified by numbers that are the power of a prime.
QUENTIN: Okay, so?
WORTH: Can you calculate that?
LEAVEN: The numbers are huge.
QUENTIN: But you can, right? You can?
LEAVEN: I'd have to calculate the numbers of factors in each set. Maybe if I had a computer.
QUENTIN: You don't need a computer.

LEAVEN: Yes, I do—Look! No one in the whole world could do it mentally!
Look at the numbers: 567 898 545. There's no way I can factor that. I can't
even start on 567. It's astronomical!
KAZAN: Two—Astronomical.

Leaven's done well up until now, but her fears are groundless. First of all,
checking for divisibility by 2, 3, and 5 is really easy, and so it's also really
easy to check for divisibility by powers of these primes. Then, the only other
prime powers Leaven needs to check are

$$7^2 = 49, \ 7^3 = 343, \ 11^2 = 121, \ 13^2 = 169, \ 17^2 = 189,$$

$$19^2 = 361, \ 23^2 = 529, \ 29^2 = 841, \ \text{and} \ 31^2 = 961 \,.$$

Given that all their lives depend upon it, this is hardly a lot of extra work.

Anyway, after Leaven makes her "astronomical" remark, they realize that
Kazan is really good at counting the factors of numbers. In fact, he gets one
wrong (the number of factors of 462), but it turns out not to matter for that
particular room.

In the end, Leaven does manage to locate the bridge. However, she winds
up dead before having a chance to escape. How typical for Hollywood: the
brainy one does all the work, and somebody else reaps the rewards.

6.7 Other Cubes

There is a sequel to *Cube*, *Cube 2: Hypercube* (2002), and a prequel, *Cube Zero*
(2004). We discuss *Cube 2* in chapter 15 ("Survival in the Fourth Dimension").
The makers of the movie *Cube 2: Hypercube* (2002) also contacted David
Pravica, but were only interested in some random formulas with which to
decorate the walls. *Cube Zero* does not feature any math.

Chapter 7
The Incredible Shrinking Room

Fermat's Room (*La Habitación de Fermat*) (2007) is a Spanish movie reminiscent of *Cube* (1997). Four "mathematicians," strangers to each other, are invited to a party by a mysterious host who calls himself Fermat. The pretext is the resolving of a great mathematical problem.

For the duration of the party, the guests are to use the pseudonyms Galois, Hilbert, Pascal, and Oliva. Away from the party, Galois is a math student who claims to have found a proof of the famous Goldbach conjecture,[1] Hilbert is an elderly mathematician, Pascal is an engineer, and all we ever learn of Oliva is that she is very good at chess.

The invitation is a trap, and the four are imprisoned in a square room. They are then confronted with mathematical problems. Whenever they fail to solve a puzzle within the specified time, the walls of the room close a distance in on them (an oldie but a goodie). They are not given much time: over one hour, the walls are set to shrink from about 7 meters in length to just 1 meter.

Fermat's Room has many cute mathematical touches: a boat called *Pythagoras*; the Kepler conjecture on the densest packing of balls; characters the same age as their mathematician namesakes were when they died (with the notable exception of Hilbert); many examples of the Goldbach conjecture; and of course the puzzles.

---

[1] The Goldbach conjecture is a very old, very famous, and still unsolved mathematical problem. It states that every even integer greater than 2 can be written as the sum of two prime numbers. That is $4 = 2 + 2, 6 = 3 + 3, 8 = 5 + 3$, and (so goes the conjecture) so on. The Goldbach conjecture also plays a central role in the murder mystery *Inspector Lewis* (2006) and in the biopic *Chen Jingrun* (2001). Jimmy Stewart also plays around with it in *No Highway in the Sky* (1951). And, in the *Futurama* episode "The Beast With a Billion Backs" (2008), Farnsworth and Wernstrom collaborate on "another elementary proof."

7.1 How Good a Puzzler Are You?

In this chapter we'll focus on the puzzles confronting the trapped mathematicians.[2] And we'll frame it as a challenge for you; see how well you would fare, if you were trapped in Fermat's deadly room.

To begin, we'll pose a couple of warmups. The first is an easy one, an old chestnut discussed by the four mathematicians while boating on a lake. You'll find the answer to this and all the puzzles at the end of the chapter.

First warmup puzzle: A man has to transport a wolf, a sheep, and a cabbage across a river in a small rowboat. On each trip, the man can only transport one passenger. When the man is not around, the wolf will eat the sheep, and/or the sheep will eat the cabbage. Devise a plan that will move all three passengers safely to the other side of the river.

The second warmup puzzle is not so easy. It was the preliminary puzzle sent to the four mathematicians, to test their worthiness of an invitation to the party.

Second warmup puzzle: What is the principle behind the ordering 8, 5, 4, 9, 1, 7, 6, 3, 2?

Warmup's over, here are the rules for our game.

- There are seven puzzles.
- Each puzzle has a time limit of five minutes (in the movie, the times allotted to the puzzles vary). Once the time limit is up, the walls begin moving. Keep track of how much you go over time. If your excess time reaches one hour, we declare you dead.
- To compensate for not having others to help you, if you're stumped by a puzzle you can choose to roll a die. If you roll a 6, we declare you dead. Otherwise, you survive, and can continue with the next puzzle.
- If your answer to a puzzle is not correct, then you have to roll the die.
- No cheating!

Are you ready to go? Good luck!

Puzzle 1: A confectioner receives three boxes of candies. One contains only mints, the second contains only chocolates, and the third contains a mixture of the two. The boxes have labels to identify the contents of the boxes. However, the confectioner was informed that all the boxes have changed labels. What is the minimum number of candies that the confectioner has to sample in order to determine the contents of all three boxes?

Puzzle 2: Decipher the following message:
00000000000000011111111100011111111110011111111110011000100
01100110001000110011111101111100111100011110001111111110000010
10101000000110101100000011111110000000000000000.

---

[2] We've slightly reworded some of the puzzles, to clarify them and to remove some ambiguities.

Puzzle 3: Inside a closed room there is a light bulb. Outside the room are three switches in the "off" position. Only one of the switches operates the light; the other two don't do anything. Once you open the door to the room, you cannot go back to move the switches. How can you determine which switch operates the light bulb?

Puzzle 4: You have a 4-minute hourglass and a 7-minute hourglass. How can you measure a period of exactly 9 minutes? The hourglasses must always be running: you cannot lay them on their sides.

Puzzle 5: A student asks his professor: "What are the ages of your three kids?" The professor answers: "If you multiply the ages you get 36, and if you add them you get my house number." "I know your house number, but that's not enough information!" says the student. To that the professor answers: "True. The oldest lives upstairs." What are the ages of the three children?

Puzzle 6: In the Land of Lies, people always lie. In the Land of Truth, people always tell the truth. A foreigner is imprisoned in a room with two doors. One leads to freedom, the other to certain death. The foreigner knows that the guardian of one of the doors is from the Land of Lies, and the guardian of the other door is from the Land of Truth. The foreigner can ask exactly one question of one of the guards. What question does he ask to find his way to safety?

Puzzle 7: A mother is 21 years older than her son. In 6 years the mother will be five times the age of her son. Where is the father right now?

7.2 Answers to the Puzzles

Answer to the First Warmup Puzzle: This is a famous puzzle. Google "sheep, wolf, cabbage."

Answer to the Second Warmup Puzzle: Consider the number names in alphabetical order: eight, five, four, nine, one, seven, six, three, two. Of course in the movie this is all done with Spanish number names: the starting arrangement is 5, 4, 2, 9, 8, 6, 7, 3, 1, corresponding to cinco, cuatro, dos, nueve, ocho, seis, siete, tres, uno.

Answer to Puzzle 1: It is important to note that all the labels have been switched, so that boxes are now definitely incorrectly labeled. Then, choosing one candy from the box labeled Mixed will suffice.

Suppose the candy chosen is a mint (we can give an exactly analogous argument if a chocolate is chosen). Then we know that it is not from the mixed box, so it must be from the mint box. Then, since the chocolates cannot be in the box labeled Chocolates, they must be in the box labeled Mints. That leaves the mixed candies in the third box, labeled Chocolates.

Answer to Puzzle 2: There are $169 = 13 \times 13$ digits in the sequence, and this is the only nontrivial way to write 169 as the product of two natural numbers. So the key is to write the sequence in 13 rows of 13 digits and interpret 0 as an empty square and 1 as a filled square. The result is a striking picture of a skull (figure 7.1), which is the required answer. In *Fermat's Room*, Galois uses the front and back of Mahjong pieces to represent the 1s and 0s.

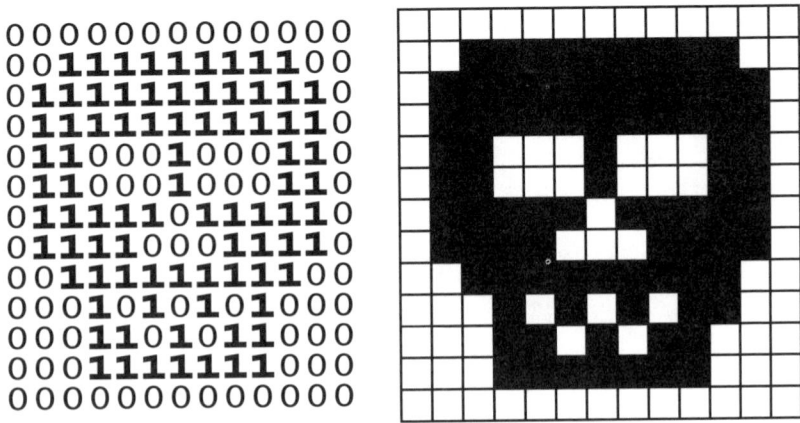

Fig. 7.1 Interpreting the string of 1s and 0s as a picture yields a . . .

This puzzle is probably inspired by the famous Arecibo message, a string of 1679 1s and 0s that was beamed to outer space, in the hope that some alien civilization would intercept it and decode it into a 23×73 pixel picture.

Answer to Puzzle 3: Label the switches 1, 2, and 3. Move switch 1 to the "on" position for a while, then switch it off again. Then move switch 2 to the "on" position, and immediately open the door. If the lamp is on, then switch 2 is connected to it. If the lamp is off and feels warm, then switch 1 is connected to it. Otherwise, switch 3 is connected to it.

In the movie it is actually not specified that we start with the switches in the "off" position. We have to require this for the solution in the movie to work.

Answer to Puzzle 4: Start both hourglasses running at the same time. As soon as the 4-minute hourglass runs out, turn it over. Then, when the 7-minute hourglass runs out, turn it over as well. At that moment, there is 1 minute left in the 4-minute hourglass. When the 4-minute hourglass runs out again (1 minute later), 8 minutes will have passed. Now turn over the 7-minute hourglass, which will run for 1 minute, giving a total of 9 minutes.

Answer to Puzzle 5: The product of the three children's ages being 36 means there are the following possibilities: (1, 1, 36), (1, 2, 18), (1, 3, 12), (1, 4, 9), (1, 6, 6), (2, 2, 9), (2, 3, 6), and (3, 3, 4). Since the student knows the house number, and since this is not enough information, two of the triples must sum to the house number. Checking the triples, the only possibility is that the house number is 13, with the ages being either (1, 6, 6) or (2, 2, 9). The final clue is that there is only one oldest child, and so the ages must be (2, 2, 9).

Answer to Puzzle 6: This is a famous puzzle, related to the liar paradox. Here is one solution. The prisoner chooses a guard, points to the door he's guarding, and asks: "If I had asked you yesterday whether you are guarding the door to freedom, would you have said 'Yes'?" It is easy to check that, no matter the guard, a "Yes" means the door is the door to freedom and a "No" means the door leads to death.

The liar puzzle also features prominently in *Labyrinth* (1986) and *The Enigma of Kaspar Hauser* (1974), and the liar paradox itself appears in *Bedazzled* (1967). In *Kaspar Hauser*, the puzzle is misstated by a pompous professor of logic, who asks Kaspar to determine whether the questioned man is a liar or truthteller: Kaspar's trivial and very funny solution is to propose asking the man whether he is a tree frog.

Answer to Puzzle 7: If the mother's age is M and the son's age is S, then

$$M = S + 21 \qquad \text{and} \qquad (M + 6) = 5(S + 6).$$

Fig. 7.2 Hilbert is figuring out the solution to Puzzle 7.

Solving for S gives $S = -\frac{3}{4}$, and so the son is negative 9 months old. So, the father is not too far away from the mother.

How did you do? Are you still alive?

7.3 Notes

There are several other movies in which puzzles play a crucial role. In *Cutting Class* (1989), a teen slasher flick, the killer sets a word problem involving trains: the answer indicates which door leads to safety. (The protagonists get it wrong.) In *Tom & Viv* (1994), Viv, T. S. Eliot's wife, is required to solve a few clever mathematical brainteasers to avoid being committed to a mental asylum. In *I.Q.* (1994), Tim Robbins has to solve some puzzles to prove that he is mathematically worthy of Meg Ryan. Other interesting puzzles are listed in chapter 17 ("Problem Corner"). In *Die Hard: With a Vengeance* (1995) Bruce Willis and Samuel Jackson need to solve a number of puzzles or face certain death. One of these is one of the oldest mathematical puzzles in recorded history:[3] "As I was going to St. Ives, I meet a man with seven wives, every wife had seven sacks, every sack had seven cats, every cat had seven kittens. Kittens, cats, sacks, and wives, how many were going to St. Ives? My phone number is 555- ... " What is the number that completes this phone number?

---

[3] A variation of this puzzle appears as Problem 79 in the Rhind Papyrus dated to around 1650 BC.

Chapter 8
Murder in the Hot House

We were just trying to put the finishing touches to this book, when we received a call from the props department of the Australian cop series *City Homicide*. In a forthcoming episode, "Hot House," they were planning on killing a couple of (fictional) mathematicians. To get things right, they wanted to talk to some (nonfictional) mathematicians. Of course, we were delighted to assist.

8.1 The Story

In "Hot House" (2010), two mathematicians are murdered. The first, Christopher Bolingbroke, is a former professor of pure mathematics at the University of Melbourne. He is now spending his time coaching brilliant math students, including his own son Harry. The second victim is Gordon Neandes, a failed mathematician. He has been "hot housing" his two brilliant children, Andrea and Liam, to have them achieve where he could not.

Harry also never made it and is currently working as a government statistician. Andrea escaped six years ago and is working as a high class prostitute. Andrea's younger brother Liam is still being tutored by Bolingbroke.

At the beginning of the show, Bolingbroke is found strapped to a chair, with his body covered by stab wounds and mathematical writing. We subsequently learn that he was forced to do the writing himself, and that the stab wounds were crossings out of his mistakes. Later, Neandes is killed in the same manner.[1]

---

[1] *Cube 2: Hypercube* (2002) also features a corpse decorated with equations. And in *Believers* (2007), the leader of a crazy cult has discovered a formula that will transport his followers to a safe place before the apocalypse; the cult members have the formula tattooed on their bodies.

It turns out that Andrea and Harry met by chance, and then planned and executed their fathers' murders, both out of revenge and to save Liam. Ironically, Liam really enjoys the math and didn't need saving.[2]

Fig. 8.1 Professor Bolingbroke, one of the murdered mathematicians.

8.2 Let's Kill Some Mathematicians

When we were first contacted by Kate and Slavko, the guys in charge of the props department of *City Homicide*, they were mainly after some writing samples for the body math, and some math-filled whiteboards and the like. In the end, we did all the displayed mathematics for the episode, checked the mathematics in the script, supplied a couple of tons of math books for props, and spent a few fascinating days on set watching the episode being created.

The script had Bolingbrook and Neandes interested in prime numbers and the Riemann hypothesis, and we chose the equations accordingly (figure 8.1). A careful look at the blackboards shows them to be mainly analytic number theory. In particular, the writing on the two bodies was borrowed from Bernhard Riemann's paper "Über die Anzahl der Primzahlen unter einer gegebenen Grösse," and from the chapter dedicated to this paper in Harold Edwards's classic book, *Riemann's Zeta Function*.[3] Not the cutting edge research called for in the script, but definitely beautiful mathematics that we hoped would look great on TV.

---

[2] A very similar plot is the basis of the episode "Bright Boy" of the TV series *Law and Order: Criminal Intent* (2006).

[3] Academic Press, New York, 1974.

Fig. 8.2 Professor Bolingbroke, with Marty on the left and Burkard on the right.

The math bodies were supposed to be prominent, and it was clear that the crew were keen to create as powerful an effect as possible. So we went on an initial trip to the Melbourne studio to practice on a trial body, and to plan it all with Kate and Slavko, the makeup department, and Kate Woods, the director.

Slavko had located some special pens that tattoo artists use for sketching their designs. They also hired a trial corpse, Luigi. He was an interesting and talkative fellow: as we drew on him, we learned of his quirky theories of love, shoes, and Cinderella.

In our trial run, we covered half of Luigi and then met with the director. After some back and forth, it was decided that we would pretty much cover all visible parts of the body with formulas. The writing on the torso was plausible. From figure 8.2, you can see how we tried to orient the writing to make it look as if Bolingbroke did it himself.

The writing on the legs down to the toes made less sense. The victim was to have his legs and left arm strapped down, with only the right arm left free to write. Bound in this way, Bolingbroke and Neandes wouldn't have been able to write so far down. However, Kate Woods had the idea of having the camera pan along the whole leg, following one line of writing (figure 8.3). The end result was quite spectacular, and certainly justified the minor fudging of reality.

We have very different handwriting, and the initial plan was to have Burkard decorate one body and Marty the other. However, during the actual filming, it was all too hectic and so we worked together on each body. Together, it took us about 45 minutes to finish a body. In any case, the writing on the skin was sufficiently indistinct for it not to matter.

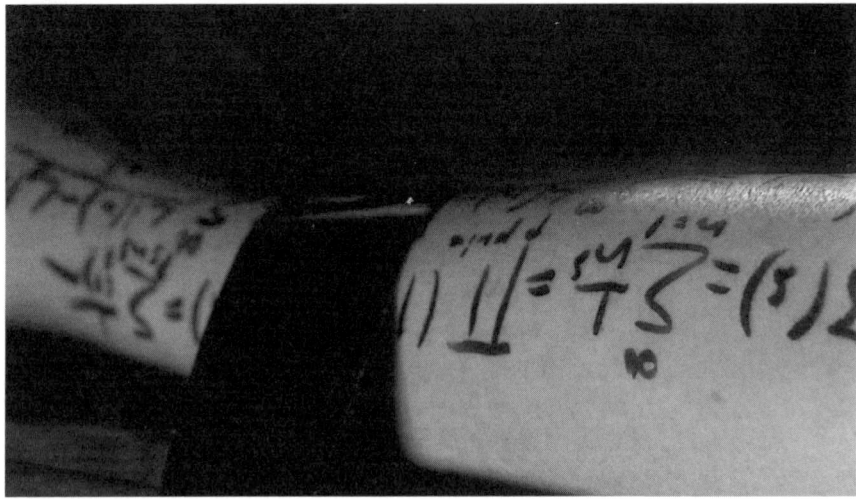

Fig. 8.3 Leonhard Euler's famous formula, relating the prime numbers to the Riemann zeta function: $\zeta(s) = \sum\limits_{n=1}^{\infty} \frac{1}{n^s} = \prod\limits_{p \text{ prime}} \frac{1}{1-p^{-s}}$.

8.3 The Writing on the Wall

Apart from decorating mathematicians, we were also very occupied filling the whiteboards and blackboards that appear in the mathematicians' offices and in Liam's room. Slavko arranged for the whiteboards to be dropped off at Burkard's home. A "few" whiteboards turned out to mean fifteen whiteboards that required filling. It took half the night (figure 8.4).

We decided to fill Liam's boards mostly with pretty calculus: a proof that

$$\int_{-\infty}^{\infty} e^{-x^2}\, \mathrm{d}x = \sqrt{\pi}\,,$$

the derivation of the reduction formula for

$$\int \sin^n x\, \mathrm{d}x\,,$$

a proof that

Fig. 8.4 Marty at work on one of the whiteboards.

$$\sum_{n=1}^{\infty} \frac{1}{n^2} = \frac{\pi^2}{6},$$

and two popular proofs that

$$\sum_{n=1}^{\infty} \frac{1}{n} = \infty.$$

One of Liam's whiteboards was a collage of some of mathematics' all-time classic formulas and one-line proofs; see figure 8.5.

8.4 Let's Run Away and Join the Circus

Altogether, we were required to be on set for three days over a period of two weeks. After decorating a body, we would stick around, just in case we were required. There was little for us to do, but it was great fun watching the filming, chatting with the cast and crew, getting in the way, but also feeling part of a great team.

We were always given the feeling that everybody was very happy to have us around, and that they were as interested in and as bemused by us as we

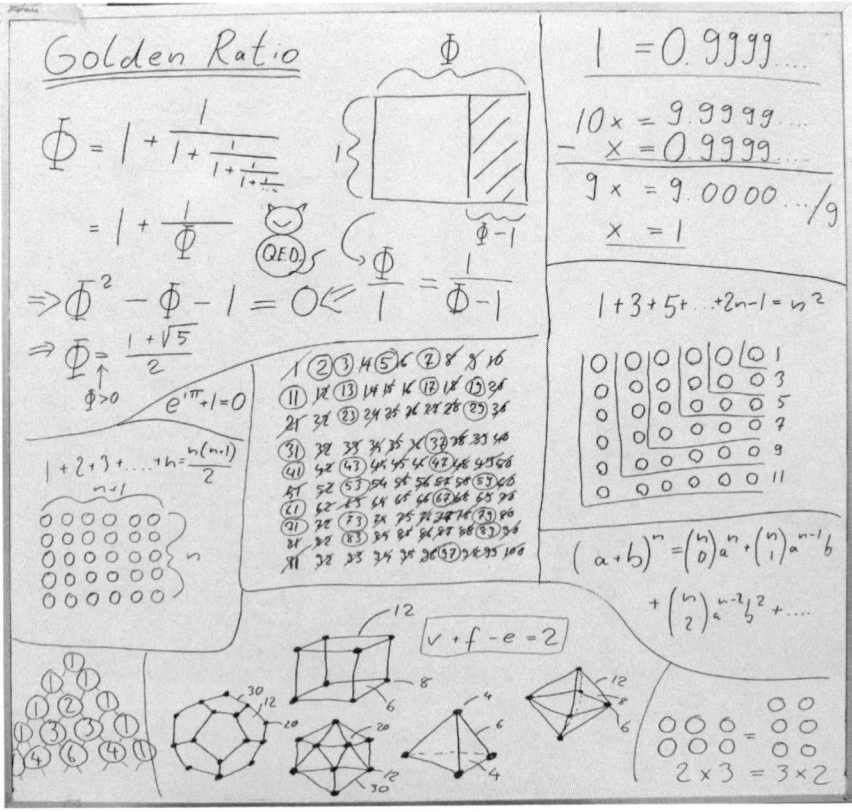

Fig. 8.5 A mathematical collage for Liam, featuring our mascot QEDcat.

were by them. It was clear that our efforts to give the episode an authentically mathematical feel were genuinely appreciated.

The lasting impression was of the general friendliness and collegiality, on everyone's part. It was all shoestring and incredibly rushed, but it was also impressively human: academia could learn a thing or two. A memorable and thoroughly enjoyable little adventure.[4]

---

[4] For many more screenshots and more detailed information, check out the page devoted to the episode on our website: www.qedcat.com/cityhomicide.html.

Chapter 9
A Word Problem for Die Hards

In *Die Hard: With a Vengeance* (1995), John (Bruce Willis) and Zeus (Samuel L. Jackson) have been set a little problem by the villain, Simon (Jeremy Irons). It involves some jugs.

0:58

SIMON [via a mobile phone]*: I trust you see the message. [The bomb] has a proximity circuit. So, please don't run.*

JOHN: Yeah, I got it. We're not gonna run. How do we turn this thing off?

SIMON: On the fountain, there should be two jugs, do you see them? A 5-gallon and a 3-gallon. Fill one of the jugs with exactly 4 gallons of water and place it on the scale and the timer will stop. You must be precise, one ounce more or less will result in detonation. If you're still alive in five minutes, we'll speak.

JOHN: Wait, wait a second. I don't get it. Do you get it?

ZEUS: No.

JOHN: Get the jugs. Obviously, we can't fill the 3-gallon jug with 4 gallons of water.

ZEUS: Obviously.

JOHN: All right. I know, here we go. We fill the 3-gallon jug exactly to the top, right?

ZEUS: Uh-huh.

JOHN: Okay, now. We pour that 3 gallons into the 5-gallon jug. Giving us exactly 3 gallons in the 5-gallon jug, right?

ZEUS: Right, then what?

JOHN: All right. We take the 3-gallon jug and fill it a third of the way ...

ZEUS: No. He said be precise. Exactly 4 gallons.

JOHN: Shit! Every cop within 50 miles is running his ass off, and I'm out here playing kids games in the park ... Look, we can't take this off, it will detonate. Just wait, wait a second. I got it! I got it! Exactly 2 gallons in here, right?

ZEUS: Right.

JOHN: Leaving exactly 1 gallon of empty space, right?
ZEUS: Yeah.
JOHN: A full 5 gallons here, right?
ZEUS: Right.
JOHN: You put 1 gallon out of 5 gallons in there, we have exactly 4 gallons
in here.
ZEUS: Yes!
JOHN: Come on. Don't spill any. Good good good. Exactly 4 gallons.
ZEUS: You did it, McClane!

We consider a general and elegant method of tackling this kind of problem.
It is based on playing billiards on a special table.

9.1 Playing Billiards with the Die Hard Problem

We use the parallelogram-shaped billiard table shown in figure 9.1.

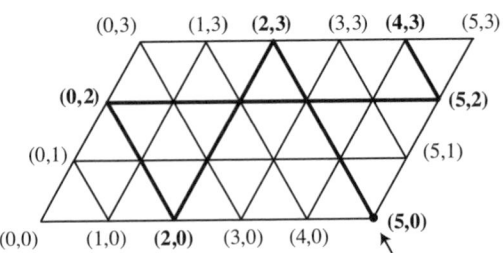

Fig. 9.1 Parallelogram billiards solves John and Zeus's jug problem.

The dimensions of the table are 5 units by 3 units, and the angles at the
corners are 60 degrees and 120 degrees, as pictured. We introduce natural
coordinates, so that the lower left corner is at $(0,0)$ and the upper right
corner is at $(5,3)$. Then we place a ball at the lower right corner $(5,0)$, and
shoot the ball as indicated. As the ball bounces off the edges of the table, we
record the bounce points:

$$(5,0) \to (2,3) \to (2,0) \to (0,2) \to (5,2) \to (4,3) \to (4,0) \to \text{etc.}$$

This sequence corresponds to John's solution of Simon's problem as follows:

(5,0): fill the 5-gallon jug;
(2,3): fill the 3-gallon jug from the 5-gallon jug, leaving 2 gallons in the
      5-gallon jug;
(2,0): empty the 3-gallon jug;
(0,2): empty the 2 gallons in the 5-gallon jug into the 3-gallon jug.
(5,2): fill the 5-gallon jug.

(4,3): fill the 3-gallon jug from the 5-gallon jug, leaving 4 gallons in the 5-gallon jug.

Following the ball further, it continues from $(4,3)$ and traverses all the inner lines on the table exactly once, before stopping at the upper left corner. The complete path is

$$(5,0) \to (2,3) \to (2,0) \to (0,2) \to (5,2) \to (4,3) \to (4,0) \to$$

$$(1,3) \to (1,0) \to (0,1) \to (5,1) \to (3,3) \to (3,0) \to (0,3) \, .$$

In particular, notice that the path of the ball includes the following points on the top and bottom edges of the table:

$$(1,0), \ (2,0), \ (3,0), \ (4,0), \ (5,0)$$

and

$$(0,3), \ (1,3), \ (2,3), \ (3,3), \ (4,3) \, .$$

Of course, 0 and 8 gallons are very easy to arrange. So, no matter the number of gallons, up to 8, that Simon might have demanded, John and Zeus could have satisfied him.

You've probably figured out why this works. The lines parallel to the x-axis represent filling or emptying the 5-gallon jug, while leaving the 3-gallon jug untouched. Similarly, the lines parallel to the y-axis represent filling or emptying the 3-gallon jug. The other lines represent pouring as much as possible from one jug into the other. So we can begin the path at $(5,0)$, and then all the indicated actions are permissible under Simon's rules.

The diagram contains a second solution to Simon's problem. Place the ball in the upper left corner and shoot in the opposite direction from before. This is the same route but traversed in the reverse direction. It gives the solution to Simon's problem as

$$(0,3) \to (3,0) \to (3,3) \to (5,1) \to (0,1) \to (1,0) \to (1,3) \to (4,0) \, .$$

9.2 A Recipe

Our first solution can be summarized as an algorithm, an example of the general recipe for solving jug problems. Beginning with both jugs empty, the key steps are as follows:

1. If the 3-gallon jug is full then empty it.
2. Pour as much water as possible from the 5-gallon jug into the 3-gallon jug.
3. If there is water left in the 5-gallon jug, then return to step 1. Otherwise
4. Fill the 5-gallon jug and return to step 2.

With this approach the water always flows as follows:

fountain >> 5-gallon jug >> 3-gallon jug >> fountain.

To obtain the second solution, in the above recipe replace every 3 by 5 and every 5 by 3. For this solution, the water flow is reversed.

9.3 Thwarting a Different Simon

What if tomorrow it's your turn? Simon hands you a 10-gallon jug and a 6-gallon jug, and you're required to obtain exactly 9 gallons of water. You could construct a 10×6 parallelogram table to try to solve the puzzle (figure 9.2).

However, even without the table it is easy to see that it is impossible to obtain a solution. Notice that 10 and 6 have a common factor of 2. So, no matter how we fill or empty the jugs, the amounts of water will only be multiples of 2. It follows that it is impossible to obtain exactly 9 gallons. Pictorially, only every second line of the billiard table is traversed.

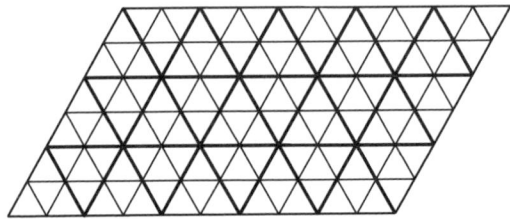

Fig. 9.2 The spacing of the path indicates the greatest common divisor of 15 and 6.

In general, the only amounts we can possibly obtain from two jugs are multiples of the greatest common divisor of the jugs. Conversely, it can be demonstrated that any (small) multiple of the greatest common divisor can indeed be obtained using the parallelogram method; for details check out the references at the end of this chapter.

This also leads to billiard tables providing a pictorial method of finding the greatest common divisor of two numbers. Construct the table corresponding to the two numbers and draw the complete path. Then the greatest common divisor is the number of spaces between the traversed lines.

9.4 The Least Common Multiple

Given two number m and n, we know that

$$\text{least common multiple of } m \text{ and } n = \frac{mn}{\text{greatest common divisor of } m \text{ and } n}.$$

So, any method for calculating greatest common divisors, such as our billiards method, also provides a method for calculating least common multiples. In fact, there is also a direct billiards method:[1]

1. Make a rectangular billiard table of length m and width n.
2. Place the ball at a corner and shoot it at 45 degrees. See figure 9.3 (where $m = 15$ and $n = 6$).

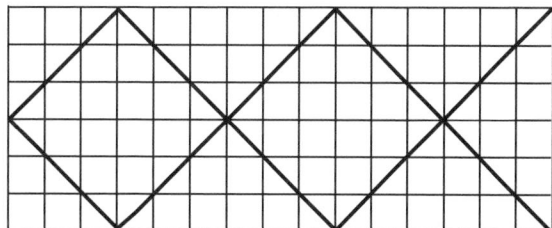

Fig. 9.3 Using a rectangular billiard table to calculate the least common multiple.

Now, count the number of times the ball travels between the left and right edges before arriving at a corner. Then the least common multiple of m and n is the product of m and this travel number. In our illustrated 15×6 example, the ball travels twice between the left and right edges: the least common multiple of 15 and 6 is $15 \times 2 = 30$.

Alternatively, the least common multiple is n times the number of trips between the top and bottom edges. In our example, this amounts to $6 \times 5 = 30$.

9.5 References

There are a number of excellent references on jug problems and on the closely related *decanting problems* or *Tartaglian measuring problems*. See, in particular:

Alex Bogomolny's Cut the Knot website. It features a very nice JAVA applet and accompanying discussion.

H. S. M. Coxeter and S. L. Greitzer, The three jug problem, *Geometry Revisited* (Mathematical Association of America, Washington, DC, 1967), section 4.6, 89–93.

P. Boldi, M. Santini, and S. Vigna, Measuring with jugs, *Theoretical Computer Science* 282 (2002), 259–270.

---

[1] A mechanical device based on this method is described in *Martin Gardner's Sixth Book of Mathematical Diversions from Scientific American* (University of Chicago Press, Chicago, 1984), page 214.

M. C. K. Tweedie, A graphical method of solving Tartaglian measuring problems, *Mathematical Gazette* 23 (1939), 278–282.

Chapter 10
$7 \times 13 = 28$

A number of Abbott and Costello's memorable routines have a mathematical flavor. The clear standout is the donut scene in *In the Navy* (1941).[1] Lou Costello plays Pomeroy, a baker, and Bud Abbott is his friend, Smokey.

1:06
SMOKEY: Hey, donuts!
POMEROY: No, Smokey! Don't!
SMOKEY: Oh, come on. One donut.
POMEROY: I haven't got enough. I can't afford it. I just baked twenty-eight of these things. Well, after all there are seven officers I've got to feed and I've just got enough to give them—thirteen apiece.

Understandably, Smokey demands a proof that 7×13 indeed amounts to 28. Pomeroy obliges, in spades. (In what follows, we indicate in bold each new step of Pomeroy's calculations.)

10.1 First Proof: Bogus Division

POMEROY: There were seven officers. There's a seven. Now I'm going to divide to prove it to you [figure 10.1]. *Now, twenty-eight donuts.*

$$7|\ \mathbf{28}\ |$$

POMEROY: Now, seven into two. You couldn't even push that big seven into that little two. Therefore we can't use the two. I'm gonna let Dizzy hold it. I'll use it later. Now, seven into eight—One.

$$7|\ 28\ |\mathbf{1}$$

---

[1] Later in this chapter we give a survey of other filmed versions of the routine.

POMEROY: Now, we're gonna carry the seven. It's getting a little heavy, so I'll put it right down there.

$$7|\ 28\ |1$$
$$\mathbf{7}$$

Fig. 10.1 $28 \div 7 = 13$.

POMEROY: Seven from eight—One.

$$7|\ 28\ |1$$
$$7$$
$$\overline{\hspace{1em}\mathbf{1}}$$

POMEROY: Now, a minute ago, we didn't use the two. I'm gonna use it now. Dizzy, give me back the two. Thanks. Put it right down there.

$$7|\ 28\ |1$$
$$7$$
$$\overline{\hspace{1em}\mathbf{21}}$$

POMEROY: Now seven into twenty-one?
SMOKEY: Three times.

$$7|\ 28\ |\mathbf{13}$$
$$7$$
$$\overline{\hspace{1em}21}$$

POMEROY: 7—28—13.
SMOKEY: Now wait a minute!

10.2 Second Proof: Bogus Multiplication

SMOKEY: Put down thirteen up there [figure 10.2]. *Now you claim that each officer gets thirteen donuts? Put down seven, draw a line. Now seven times thirteen is what?*

$$
\begin{array}{r}
13 \\
7 \\
\hline
\end{array}
$$

Fig. 10.2 $13 \times 7 = 28$.

POMEROY: Twenty-eight.
SMOKEY: Prove it.
POMEROY: Seven times three?
SMOKEY: Twenty-one.

$$
\begin{array}{r}
13 \\
7 \\
\hline
21
\end{array}
$$

POMEROY: Seven times one.
SMOKEY: Seven.

$$
\begin{array}{r}
13 \\
7 \\
\hline
21 \\
7 \\
\hline
\end{array}
$$

POMEROY: Seven and one?
SMOKEY: Eight.

$$
\begin{array}{r}
13 \\
7 \\
\hline
21 \\
7 \\
\hline
\mathbf{8}
\end{array}
$$

POMEROY: Two.

$$
\begin{array}{r}
13 \\
7 \\
\hline
21 \\
7 \\
\hline
\mathbf{28}
\end{array}
$$

10.3 Third Proof: Bogus Addition

SMOKEY: Oh no. Come on. We add this up. Put down thirteen seven times [figure 10.3].

Fig. 10.3 $13 + 13 + 13 + 13 + 13 + 13 + 13 = 28$.

SMOKEY: Now we're getting it. You claim all this added up amounts to what?
POMEROY: Twenty-eight.
SMOKEY [adding up the 3s from the bottom to the top]: *3, 6, 9, 12, 15, 18, 21, ...*
POMEROY [takes over, adding the 1s from the top to the bottom]: *22, 23, 24, 25, 26, 27, 28!*

10.4 Play it Again, Abbott

As for most of their routines, Abbott and Costello employed the $7 \times 13 = 28$
skit a number of times. They did an excellent version in *Little Giant* (1946).
On TV, it appears in *The Colgate Comedy Hour* (1952) and in *The Abbott
and Costello Show* (1953).[2]

The routine actually long predates Abbott and Costello. In his book *Knot-
ted Donuts and Other Mathematical Entertainments*,[3] Martin Gardner traces
it back to Irvin S. Cobb's 1923 anthology of jokes, *A Laugh a Day Keeps the
Doctor Away*. It is possible that Flournoy Miller and Aubrey Lyles performed
the routine in the 1921 Broadway show *Shuffle Along*;[4] if so, they almost certainly performed it much earlier on the vaudeville stage.

Probably the earliest filmed version appears in the short film *Jimtown
Speakeasy* (1928), in which Miller and Lyles prove $3 \times 17 = 24$.[5] In 1951,
Miller and Ches Davis performed the $7 \times 13 = 28$ version in *Yes Sir, Mr.
Bones*. In the same year, Ma and Pa Kettle proved $5 \times 14 = 25$, in *Ma and Pa
Kettle Back on the Farm*. Finally, Flip Wilson and a young Michael Jackson
did a fine job with $7 \times 13 = 28$ on *The Flip Wilson Show* (1972).[6]

Fig. 10.4 Michael Jackson performing the routine on the *Flip Wilson Show*.

---

[2] Episodes 2.36 and 1.8, respectively.

[3] W. H. Freeman & Company, London, 1984.

[4] The claim is made by Mel Watkins in his book *On the Real Side* (Simon and Schuster,
New York, 1994), 163, with a less precise claim appearing in Ronald Smith's *Comedy Stars
at 78 RPM*, (McFarland, 1998), 149. However, neither book cites a reference supporting the
claim. We are also unaware of a contemporaneous theatre review that mentions the routine.
A discussion of the matter with David Thompson of Agnes Scott College indicates that
there is definitely some reason to doubt the claim. David and his colleagues Lyn Schenbeck
and Constance Hill have produced a critical edition of the script of *Shuffle Along*, based
upon what are very likely 1921 working copies, one the former property of Flournoy Miller.
The scripts and the accompanying annotations give no indication of the routine.

[5] A loosely related routine appeared soon after, in the Amos 'n Andy movie *Check and
Double Check* (1930). This skit is much simpler but still quite funny.

[6] Episode 3.4.

10.5 General Bogus Math

Martin Gardner discusses the question originally posed and answered by mathematician William R. Ransom:[7] Is there something special about the numbers 7, 13, and 28 that makes this routine work? Clearly not, as demonstrated by the routines of Lyles and Miller and Ma and Pa Kettle. Exactly which combinations of numbers work?

To tackle this question, let's write the three numbers as n, lr, and LR. Here, l and L indicate left digits, and r and R indicate right digits. So, for example, the routine $7 \times 13 = 28$ arises from choosing $n = 7$, $l = 1$, $r = 3$, $L = 2$, and $R = 8$.

Abbott's bogus multiplication then takes the form $n \times lr = LR$, and his calculation becomes

$$
\begin{array}{c}
13 \\
7 \\
\hline
21 \\
7 \\
\hline
28
\end{array}
\qquad \longrightarrow \qquad
\begin{array}{c}
lr \\
n \\
\hline
n \cdot r \\
n \cdot l \\
\hline
LR
\end{array}
$$

So in order for the bogus multiplication to work we require that

$$ n \cdot (r + l) = LR . $$

A little calculation then shows that this equation also suffices for the bogus addition and bogus division. However, for the division routine to work neatly in the way Abbott performs it, we also require that n goes into R at least l times, but not $l + 1$ times. That means, we also require

$$ n \cdot l \leqslant R < n \cdot (l + 1) . $$

With these two observations, it is not hard to work through all the possible candidates. Consider $n = 7$, for example. Since $7 \cdot l \leqslant R$, we must have $l = 1$, and the only possibilities for R are 7, 8, and 9. The only multiplications by 7 giving such last digits R are $7 \times 4 = 28$ and $7 \times 7 = 49$. These give us the two bogus multiplications $7 \times 13 = 28$ and $7 \times 16 = 49$.

The following is the complete multiplication table:

$2 \times 15 = 12$	$3 \times 14 = 15$	$4 \times 13 = 16$	$5 \times 12 = 15$	$6 \times 12 = 18$	$7 \times 13 = 28$	$8 \times 15 = 48$
$2 \times 25 = 14$	$3 \times 17 = 24$	$4 \times 15 = 24$	$5 \times 14 = 25$	$6 \times 15 = 36$	$7 \times 16 = 49$	
$2 \times 35 = 16$	$3 \times 24 = 18$	$4 \times 18 = 36$	$5 \times 16 = 35$	$6 \times 17 = 48$		
$2 \times 45 = 18$	$3 \times 27 = 27$	$4 \times 25 = 28$	$5 \times 18 = 45$			

---

[7] See also the article by J. Jaroma and A. Kumar, Ma and Pa Kettle Arithmetic, *Journal of Recreational Mathematics* 33 (2004–2005), 22–28.

Chapter 11
One Mirror Has Two Faces, Two Mirrors Have . . .

The Mirror Has Two Faces (1996) is an unusual story for Hollywood: boy meets girl, boy and girl have trouble, and eventually both live happily ever after.

Okay, maybe not that unusual. However, in this case the boy is math professor Gregory Larkin (Jeff Bridges). Greg is seeking a simple and sex-free relationship. The lucky girl is Rose Morgan (Barbra Streisand), a professor of literature and keen on romance.

The movie is full of mathematics: conversations about prime numbers, calculus lectures, and eye-catching blackboards. If you're a typical mathematician, surviving in a bewildering, unmathematical world, it is the movie for you.

Watch this movie together with friends and impress them by identifying the obscure mathematical tidbits and by pointing out the crazy bloopers. Maybe borrow some mathematical-romantic lines to impress a nonmathematical love interest. Or, if you're a teacher, you might try Rose's advice on how to add zing to your classes. This chapter can be your guide.

11.1 Real Life

Nothing beats living in a purely mathematical world, as Greg sums up admirably at his book launch:

0:04
GREG: As I stand here at the end of the journey, I am reminded of something Descartes once said: For whether I am awake or asleep, two and three will always make five, the square can never have more than four sides, and it does not seem possible that truths so clear and apparent can be suspected of any uncertainty.

That is all well and good, but what about real life? Felicia, the expert on the sex hotline that Greg consults, has some sound advice:

0:10

FELICIA: Life is very complex, there are no guarantees.
GREG: Why should that be? The mathematical world is completely rational,
completely uncomplicated by sex.
FELICIA: You think too much, Honey.

11.2 Prime Numbers

If you're hunting for dinner conversation that is simultaneously mathematical
and romantic, you may want to go for the twin prime conjecture:

0:30

GREG: It is interesting how coupling occurs throughout nature and in math-
ematics.
ROSE: Oh yeah, you were telling me something about pairs.
GREG: Oh, the twin prime conjecture. Yes, well, it explores pairs of prime
numbers, numbers that are only divisible by themselves: 3-5, 5-7, not 7-9,
because ...
ROSE: ... 9 can be divided by 3.
GREG: That is right. That's right. Then you have 11-13, 17-19, and so on,
and what was discovered was that what often occurred were pairs that were
separated by ...
ROSE: ... one number in between.
GREG: Exactly, exactly.
ROSE: This twin prime conjecture is interesting. What would happen if you
counted past a million. Would there still be pairs like that?
GREG: I can't believe you thought of that. This is exactly what is yet to be
proven in the twin prime conjecture.

Rose has a great idea for a present for a mathematician boyfriend:

0:51

GREG: What are these? Dice?
ROSE: They once were dice, but now they're cufflinks with ... [figure 11.1]
GREG: Prime numbers!
ROSE: 2, 3, 5, no 9.

And don't miss out on any opportunity to make that special moment mathematical:

1:32

GREG: My number here is 01712577355 ... All prime numbers, by the way.

At this stage, Rose isn't talking to Greg. Things are looking bad. So it's
understandable that a distracted Greg totally forgets that 0 and 1 are not
prime numbers.

As Greg becomes more distressed, he takes his frustration out on his students:

Fig. 11.1 Prime number cufflinks—the perfect present for that mathematician boyfriend.

1:45
GREG: D [the grade assigned to the student's paper]. *Congratulations, you're improving.*
STUDENT: I still don't understand what you're saying about twin primes.
GREG: I'm explaining it to you.
STUDENT: But I still don't understand.

It's probably not that hard, but Greg's explanation is very unlikely to help:

GREG: Don't you know that it is possible to remove an infinite number of elements from an infinite set and still have an infinite number of elements left over. We spent quite a bit of time on this already. Christ, my wife understood this on our first date!

Greg is right, of course. For example, removing all the even numbers from the natural numbers still leaves the infinitely many odd numbers. However, this observation about infinity doesn't tell you a whole lot about twin primes.

11.3 Calculus

This movie is a great source of calculus clips. In the very first scene, Greg is having a blast proving a theorem. Of course, his students are bored to tears.

0:01
GREG: $f(x) = e^y$ times e^x, which is what we were trying to prove. Notice the elegance of the proof. It's beautiful. It actually reminds me of a quote by Socrates: If measure and symmetry are absent from any composition in any degree, it ruins both the ingredients and the composition. Measure and symmetry are beauty and virtue the world over.

Flowery language, but it is indeed an elegant proof. We've reconstructed the blackboards, the numbers indicating the order to follow (figure 11.2). Greg is proving the law for the product of exponential functions (blackboard 4), an

Board 1:

$$\Rightarrow \frac{d \ln y}{dy} = \frac{1}{y} \qquad (1)$$

$$\frac{de^x}{dx} = e^x \Rightarrow y = e^x$$

Board 2:

Exponential function e^x $\qquad (2)$

<u>Definition 1:</u> e^x is the inverse function of $x = \ln y$

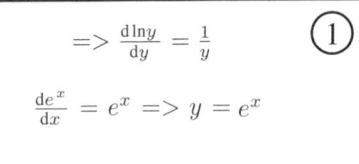

Board 3:

<u>Definition 2:</u> e^x is the unique solution of the differential equation $\qquad (3)$

$$\frac{Df}{dt} = f \quad \text{initial condition } f(0) = 1$$

(use the existence and uniqueness theorem of ordinary differential equations)

Board 4:

$\qquad (4)$

Good to show the law of exponents *from this* definition

law of exponents: $\boxed{e^{x+y} = e^x e^y}$

Board 5:

Proof: Set $\boxed{f(x) = e^{x+y}}$ $\qquad (5)$

$$y = b \quad \text{a constant}$$

Compute $\frac{df}{dx}$ by the <u>chain rule</u>

$$\frac{df}{dx} = e^{x+b} \frac{d(x+b)}{dx} = e^{x+b} = f(x)$$

Board 6:

$\Rightarrow f(x)$ is a solution to the ODE $\frac{Df}{dt} = f$ with initial value $f(0) = e^{0+b} = e^b$. Now set $b = y$

$$\Rightarrow \quad \boxed{e^{x+y} = e^x e^y} \qquad (6)$$

Fig. 11.2 The first wall of blackboards. Greg begins with the top equation of board 5.

"aside" to the second definition of the exponential function (blackboard 3). The proof itself is an application of the chain rule for differentiation (blackboards 5 and 6).

In this scene, we also catch a glimpse of Greg's textbook: *Calculus—A Computer Algebra Approach* by I. Anshel and D. Goldfeld.[1] The textbook also appears later, when Greg is talking about implicit differentiation:

0:27
GREG: Implicit differentiation is often thought of as complex, but in reality it is quite simple. In reality, it's simply a matter of pretending that y is a function of x and so ...

Fig. 11.3 Differentiating the symmetries of an equilateral triangle?

At this point, we have an elephant-sized blooper. Greg's words are fine, but he is also pointing to a blackboard detailing the symmetries of equilateral triangles (figure 11.3): absolutely nothing to do with implicit differentiation.

In a later scene, Greg actually applies implicit differentiation:

0:38
GREG: x squared plus y squared equals 16. How do we find $\frac{dy}{dx}$ as an implicit function of x and y? Well y is a function of x, so we differentiate both sides.

---

[1] Perhaps an early example of mathematical product placement.

The left-hand side is pretty easy, what is the left-hand side? The left-hand side is 2y.

Fig. 11.4 Greg demonstrates his brilliant lecturing technique.

Greg has just finished writing the "x squared." The chain rule is written as $\frac{df}{dx} = \frac{df}{dy}\frac{dy}{dx}$, and Greg seems to have differentiated the function $f(x) = \cos(x^2 + 2x)$ (figure 11.4). Then, setting $y = x^2 + 2x$, we obtain

$$\frac{df}{dy} = -\sin(y) \qquad \text{and} \qquad \frac{dy}{dx} = 2x + 2.$$

Greg is a boring lecturer, and he knows it. He asks for help from Rose, who is a charismatic humanities lecturer. This leads to a couple of scenes featuring the simple math of constant acceleration.

0:39
ROSE: Come on, teach me something, anything.
GREG: Okay, if a ball is thrown into the air and its height $h = 100t - 16t^2$.
ROSE: Gregory you see, what are you doing? Turn around, turn around, talk to me!
GREG: Okay, t is time in seconds. What limit is the speed approaching when t approaches . . .
ROSE: You lost me. Put it in some context. Make up a story, jazz it up a little bit, find some humor in it.
GREG: Humor in calculus?
ROSE: Hmm. Well, try telling a story.

Later, in class, Greg returns to this topic. On one of the blackboards is written $\frac{d^2 f}{dx^2} = c$, and the other contains standard formulas for motion under constant acceleration.

GREG: . . . equation such as the second derivative of the function f with respect to the variable x equals a constant . . .

A student yawns, prompting Greg to try Rose's suggestion:

GREG: Anybody see the game yesterday? That Marrakesh, what a bum, hey? Let me try putting this another way. When measuring trajectories, if a batter hits a ball, how can we determine how far that ball will go? What are the variables needed to hit a home run, pretending for a moment that the bum could hit a home run.
STUDENT: The velocity with which the ball leaves the bat.
GREG: The velocity, correct.

Actually, a math lecturer can immediately recognize the above conversation to be a bit of a blooper. It is highly improbable that Greg's spooned-on relevance would actually entice the bored students. Especially so, since Greg's heart is clearly not in it: like many pure mathematicians (pun intended), Greg is uninterested in baseball. However, Rose eventually makes him see the light:

1:06
GREG: I don't see the point of playing a game where you wind up at the same spot where you started.
ROSE: Well, actually this should interest you because it's about stats and averages.
GREG: Stats and averages? Explain that to me.
ROSE: Well, see, every time a player comes up, they flash this three-digit number and that is the player's average. That is, how many times he hits the ball in ratio to how many times he comes up to bat.

If you watch the movie carefully, you'll notice many more calculus fragments on the blackboards: sine curves, regions under functions ripe for integration, and so on.

11.4 Mathematical Miscellany

Pointing out the above details should be sufficient to engage nonmathematical companions. However, to impress mathematicians, more is definitely required. To that end, we now offer a survey of the beautiful mathematics lurking in the background of the movie (almost none of which has an ounce of relevance to Greg's calculus course).

Be warned, there is some high-level mathematics peeking out of those black-boards. So, don't worry too much about the details. When need be, simply admire the graphics, and record to memory those very impressive buzzwords.

0:16
The two diagrams on the bottom right blackboard in figure 11.5 show the simplest knot, the *trefoil knot*. illustrated as a *torus knot*. The middle black-boards appear to be *Coxeter-Dynkin diagrams*, indicating the singularities of

Fig. 11.5 Newton's diverging parabolas, Dynkin diagrams and the torus knot.

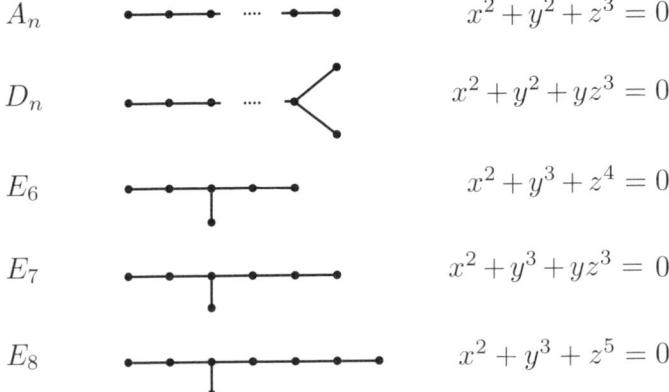

Fig. 11.6 Greg's Coxeter-Dynkin diagrams.

the algebraic curves given by the adjacent equations. (The reconstruction in figure 11.6 contains a bit of guesswork.)

The left blackboards appear to show two examples of *Newton's diverging parabolas*. The bottom curve with the cusp, $y^2 = x^3$, is the so-called *semicubial parabola*, or *Neile's parabola*. The other curve, $y^2 = (x^2 - 1)(x - a)$, is one of *Newton's egg curves*.

0:27

Rose is in front of a bulletin board, to which is pinned a very pretty graphic (figure 11.7). The original can be found on the website for the software application *Quasitiler*. As is explained there, the *Penrose tiling* that forms the background of the picture is obtained by "projecting a slice of a regular lattice in five-space onto three-space, and the one-skeleton of a five-cube projected onto three-space." Boy!

Fig. 11.7 A Penrose tiling.

1:07

The camera swings across a *Möbius strip*, a "dodecahedral space," the Dynkin diagrams again, to the Newton curves (which have been changed); see figure 11.8.

Fig. 11.8 The net of a dodecahedron, a Möbius strip, and some new Newton curves.

1:45
In the final classroom scene, we see some blackboards describing a predator-prey model. This material can be found in most introductory texts on differential equations and fits in well enough with Greg's calculus lectures. However, the blackboards in the back of the room are much more exotic.

Fig. 11.9 More beautiful and completely arbitrary mathematics to decorate the movie.

From right to left, the blackboards feature an alternating knot (of course immediately recognizable as the 7_6 knot in Tait's classical list!); the Borromean rings; Sierpinski's carpet fractal; the graph of a damped oscillation; and some spiral thingy (figure 11.9). All beautiful mathematics, and all pretty much irrelevant to a subject on calculus and differential equations.

11.5 The Other Mirror

The Mirror Has Two Faces is very loosely based on the 1956 movie with the same title. This movie is less of a Hollywood story: boy meets girl, boy and girl marry, girl gets a face lift, and boy kills plastic surgeon. Like Greg, the French boy is a teacher of mathematics. However, there is only one mathematics scene, in which he is relentlessly grilling one of his students (figure 11.10).

0:50
STUDENT [reading off the blackboard]*: "What is the smallest number containing three digits?"* [Student writes 000.]
TEACHER: No, Monsieur Biniole. [While he is waiting he is slapping a ruler into his palm. The student writes 111.]
TEACHER: No, Monsieur Biniole. [The student scratches his bottom.] *It's no use scratching like that. You won't find the answer in the seat of your pants.*

Fig. 11.10 The smallest number containing three digits?

Actually, 000 seemed like a very reasonable answer to us. Unfortunately, just as Monsieur Biniole starts to write again, the scene ends. It seems we're fated to never know what the smallest three-digit number really is.

11.6 Notes

The math consultant for *A Mirror Has Two Faces* was the mathematician Henry Pinkham from Columbia University. In an interesting article in the *New York Times*, Pinkham recalled his experience coaching Jeff Bridges. It seems Bridges "went to a lot of effort to learn lines that would be convincing from a mathematical point of view." Apparently, Bridges even adopted a few of Pinkham's mannerisms. However, Pinkham also made it clear that the boring lecturing was not modeled on him. He assured the readers that his own teaching "is at a higher level of interest," and that the math professors at Columbia "know how to tell a story."

Chapter 12
It's My Turn for Some Serious Mathematics

To our knowledge, *It's My Turn* (1980) is the only movie with a scene dedicated to the detailed proof of a mathematical theorem.[1] It is astonishing to have such a scene begin a Hollywood movie. Perhaps even more astonishing, the proof is correct.

This movie is also one of the few to mention the mathematical field of group theory.[2] It is the only movie we know of that refers to the classification of the finite simple groups, one of mathematics' holy grails.

The math consultant for the movie was Benedict H. Gross, a professor at Harvard University. To understand more of how the movie came to be, we contacted Professor Gross, and he kindly responded to our many questions. We have incorporated a number of his very informative replies.

Please be warned, there is some high-level mathematics on the road ahead. If you are not familiar with this mathematics, then we recommend you simply marvel at it, as was intended, and just enjoy the ride. In any case, please buckle your mathematical seat belts!

12.1 The Snake Rears Its Lovely Head

The main character is Kate Gunzinger (Jill Clayburgh), a math professor in Chicago. The theorem she is proving is the famous *snake lemma* from homological algebra:

---

[1] Mention must be made of the episode "The Prisoner of Benda" (2010) of the TV series *Futurama*. This terrific episode involves the switching of minds between pairs of bodies, which then cannot be directly switched back. At the end of the episode, the Harlem Globetrotters declare that all the minds can be returned to their original bodies, with the assistance of two new bodies. A full proof, specifically worked out for the episode, is displayed in the background.

[2] There is some factoring of Lie groups in the background of *Codename Icarus* (1981). And, in case it counts, reference is made to "isomorphic group therapy" in the episode "Trees Made of Glass, Part 2" (2005) of the TV series *Threshold*.

Snake Lemma: *Suppose we have the commutative diagram*

$$0 \longrightarrow A \xrightarrow{f} B \xrightarrow{g} C \longrightarrow 0$$
$$\downarrow\alpha \quad \downarrow\beta \quad \downarrow\gamma$$
$$0 \longrightarrow A' \xrightarrow{f'} B' \xrightarrow{g'} C' \longrightarrow 0$$

in some Abelian category. Then there is a long exact sequence

$$0 \to \ker\alpha \to \ker\beta \to \ker\gamma \to \operatorname{coker}\alpha \to \operatorname{coker}\beta \to \operatorname{coker}\gamma \to 0.$$

Of course, if you are unfamiliar with all the jargon then this won't make a bit of sense. And this is neither the time nor the book for a lesson on homological algebra. But as a hint to what this is all about, the capital letters in the diagram stand for mathematical worlds of the same type (such as vector spaces or groups), and the arrows stand for special functions between these worlds (homomorphisms).

It is the first scene of the movie, and on Kate's blackboard the snake lemma takes the form shown in figure 12.1.

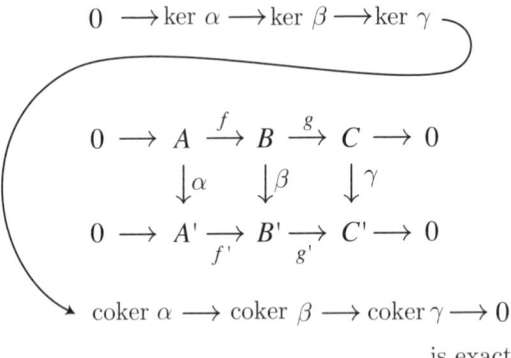

Fig. 12.1 Kate's rendering of the snake lemma.

At the beginning of the scene, the whole diagram is already visible on the blackboard, except for the long curved arrow; see figure 12.2. Kate must prove the existence of the special function corresponding to this long arrow. Having done so, she adds the arrow, completing the diagram.

The designation "snake" obviously refers to the form of this arrow, and there is good reason for representing the arrow in this manner. The object ker α is associated with A, coker α is associated with A', and so on. Then the

Fig. 12.2 Kate proving the snake lemma.

arrow joins the right side of the top row to the left side of the bottom row of the commutative diagram, and so naturally takes the shape of a snake.[3]

Comparing the two screenshots above, we can detect a small blooper: the g in the right shot has been rewritten as a q in the left shot. This shows that this scene took more than one take, and one can see why. In fact, according to Gross, the scene "took quite a number of takes. Clayburgh memorized the mathematics like you or I might memorize a greeting in Chinese, syllable by syllable." It has to be said, she did a fine job:

0:01

KATE: Let me just show you how to construct the map s, which is the fun of the lemma anyhow, okay? [This map s corresponds to the long curved arrow.] *So you assume you have an element in the kernel of γ, that is, an element in C, such that γ takes you to 0 in C'. You pull it back to B, via map g, which is surjective, ...*

COOPERMAN [the student from hell]*: Hold it, hold it. That's—that's not unique.*

KATE: Yes, it is unique, Mr. Cooperman, up to an element in the image of f, all right? So we've pulled it back to a fixed b here and then you take β of b, which takes you to 0 in C', by the commutivity [should be commutativity] *of the diagram. It is therefore in the kernel of the map g', hence is an image of f' by the exactness of the lower sequence ...*

COOPERMAN: No.

KATE: ... so we can pull it back to an element in A' ...

COOPERMAN: It's not well defined.

KATE: ... which it turns out is well defined modulo the image of α, and thus defines an element in the cokernel of α [at this point she draws the snake

---

[3] The snake lemma is just one of a number of catchily named lemmas concerning commutative diagrams. There are also the four lemma, the five lemma, and the 3×3 lemma, to name just a few. As with the snake lemma, the names are suggestive of the associated commutative diagrams.

arrow], *and that's the snake. And on Monday we'll address ourselves to the cohomology of groups* [Cooperman raises his hand], *and Mr. Cooperman's next objections.*

At the end of the scene, Kate raises the blackboard to reveal another blackboard underneath. It contains three basic statements about sequences of maps in homological algebra (figure 12.3).

$$0 \longrightarrow A \xrightarrow{f} B \qquad \text{is exact iff } f \text{ is injective}$$

$$B \xrightarrow{f} C \longrightarrow 0 \qquad \text{is exact iff } f \text{ is surjective}$$

$$0 \longrightarrow A \xrightarrow{f} B \longrightarrow 0 \qquad \text{is exact iff } f \text{ is an isomorphism (bijective)}$$

Fig. 12.3 Basic homological algebra.

There are a couple more incomplete blackboards visible in the scene. Only one can be reconstructed with any confidence (figure 12.4). According to Gross, it was probably "intended as a homework problem on homological algebra."

If

$$0 \longrightarrow A \longrightarrow B \longrightarrow C \longrightarrow 0 \text{ is exact, then}$$

$$0 \longrightarrow \text{Hom}(Q,A) \longrightarrow \text{Hom}(Q,B) \longrightarrow \text{Hom}(Q,C) \text{ is exact.}$$

How to construct the maps in the 2nd sequence and prove exactness?

Fig. 12.4 Homological homework.

12.2 The Classification of Finite Simple Groups

The classification of finite simple groups, completed around 1981, was one of the crowning achievements of twentieth century mathematics. The proof, the joint effort of over a hundred mathematicians, is comprised of over 15,000 pages in over 500 journal articles, published between the late 1940s and the early 1980s. The classification theorem states the following:

Every finite simple group is isomorphic to a group of prime order, an alternating group A_n, one of the finite groups of Lie type, or one of 26 sporadic simple groups.

The sporadic simple groups are some of the weirdest and most wonderful creatures in the whole of mathematics, arguably the most bizarre of which is the humongous *monster group*. Five of these sporadic groups had been known since the middle of the nineteenth century. The remaining twenty-one, including the Monster, were only discovered between 1965 and 1975.[4]

Gross remarks upon this aspect of the movie: "When I first read Bergstein's script,[5] she had Kate working on the theory of Abelian groups. I told her that this was not of much interest at the moment,[6] and that she should move her into the classification of the sporadic simple groups. There were many other mathematical impossibilities, and I was amazed when she told me that the movie was going into production in two weeks."

Directly after the snake lemma scene, reference is made to Kate working on groups.

COOPERMAN: This stuff is just garbage. That's another diagram chase. When are we going to move on to something interesting, like your new group. Any progress with the 2-fusion?
KATE: No, still stuck.
COOPERMAN: Maybe you have gone as far as you can with it Dr. G.
KATE: That's possible.
COOPERMAN: I've started looking at it with a whole new angle.
KATE: Oh?
COOPERMAN: If it works, I'll be famous.
KATE: Oh, that would be terrific. I can relax. I'll be famous for having taught you.

0:07
Spurred into action by Cooperman's dig at her, Kate begins calculating on the back of an envelope. Homer, her clueless boyfriend, is keeping her company.

KATE: I'd like to kill that little Cooperman. Now, he is working on the 2-fusion, Jesus. You know what I'd like?
HOMER: What?
KATE: If I could just solve this problem. You understand, I would be in a class with Euclid and Newton, really, I would be—except for Newton made his breakthrough when he was 22.

Here is Gross on the mathematics of this scene: "I forget (what is written on this envelope), although the diagram on top is probably a modified Dynkin diagram, much used in the study of sporadic finite groups. You can find similar diagrams in the Atlas.[7] I think the series was supposed to be a Thompson

---

[4] For a popular account of the classification, see Markus du Sautoy's book, *Symmetry: A Journey into the Patterns of Nature* (Harper, 2008).

[5] Eleanor Bergstein was the screenwriter for *It's My Turn*.

[6] That is an understatement.

[7] J. H. Conway, R. T. Curtis, S. P. Norton, R. A. Parker, and R. A. Wilson, *Atlas of Finite groups: Maximal Subgroups and Ordinary Characters for Simple Groups* (Oxford

series, giving the character values of an infinite sequence of representations on a fixed conjugacy class. These turn out to be modular forms for certain sporadic groups, like the Monster. It was all new mathematics at the time of the filming."

0:17

A conversation with the professors interviewing Kate for a new position.

FIRST PROFESSOR: Jeremy Grant at Yale. He is a great fan of your thesis. Have you done any new work on your group yet?
KATE: No, not yet.
FIRST PROFESSOR: I broke my back on group theory. It has moved way past me now. You younger minds have to take over.
SECOND PROFESSOR: Has your work come to a standstill, Dr. Gunzinger?
KATE: I certainly hope not.

0:20

Kate is at dinner with her father and others.

FATHER: How did the interview go?
KATE: I should have kept my mouth shut. I don't think I'm going to get it.
FATHER: That doesn't sound like you. This is the girl who got one hundred on every math exam except for 98 in plane geometry. Her thesis was on sporadic groups.
GUEST: Tell me, what did you get wrong in plane geometry?
KATE: The problem was to compute the area of a patio around a pool. I applied the right method, but I put the patio inside the pool.

1:25

Back on her home campus, Kate runs into Cooperman.

COOPERMAN: Did you make any progress?
KATE: Well, I tell you I have been thinking. I think, I have been looking in the wrong place. I have some new ideas about the 2-fusion.
COOPERMAN: What? Why? Do you mean dot o to dot g?
KATE: Right, just in its simplest case, though.
COOPERMAN: But that might be the hinge of the whole problem. Yeah, right away that's going to give you the quotient, that's immediate.
KATE: It's just a beginning, though.
COOPERMAN: Show me what you are talking about.
KATE: I can't do it now.
COOPERMAN: Show me.
KATE: I'm going to be here for a while.
COOPERMAN: The classification might even drop right out. This is incredible. If this works, we could be famous.
KATE: Listen, it's just the beginning. The tough part is working it out.

University Press, Oxford, 1985). This is a very unusual A3-sized book. A must-see for any true math enthusiast.

We never find out whether they solve the problem.

12.3 Questions and Answers

We did our best to exhaust Professor Gross with our many questions on how the movie math came to be. The following is some of our conversation.

Question: In the movie it is not quite clear what Kate is actually trying to do. On the one hand, people refer to "her group" which seems to indicate that she already found a sporadic finite simple group as part of her dissertation. On the other hand, she is still trying to do something with it. Clearly, this was not really developed in the movie. Do you still remember what people had in mind here? And what about this 2-fusion business?

Answer: They didn't care that she has a serious research problem. Just that the language was impressive and essentially correct.

Question: How did you get involved in the making of this movie?

Answer: I was an instructor at Princeton, where Bergstein was living at the time. She asked the department chair, Hale Trotter, if she could visit the class of a young mathematician in the department, to model a character (played by John Shea, but cut from the final version). After coming to class, she took me to lunch and showed me the script. When I came across the Abelian group stuff, I suggested that she needed some help, and that got me involved. My main contribution was the opening scene on the snake lemma.

Question: Did you or any other real mathematician by any chance make an appearance in the movie?

Answer: Not that I know of. I tried to keep my distance, and asked not to be in the credits.

Question: Was that your handwriting on the blackboards?

Answer: I can't remember. Probably. I was flown out to Hollywood for the filming of the math scenes.

Question: Are there any other interesting anecdotes that come to mind?

Answer: When I protested at the line that Clayburgh used about being over twenty, therefore incapable of doing math, and suggested that it be modified to being over thirty, I was told that it was impossible for Clayburgh to admit in a movie that she was over thirty. Otherwise she couldn't get leading parts. [*This scene did not make it into the movie*].

12.4 Notes

The Movie in the Math: The snake lemma scene is mentioned in some advanced math texts, such as C. A. Weibel, *An Introduction to Homologi-*

cal Algebra (Cambridge University Press, Cambridge, 1995) and C. L. Scho-
chet, The topological snake lemma and corona algebras, *New York Journal
of Mathematics* 5 (1999), 131–137.

Commutative Diagrams in Movies: There are two other movies we know
of that feature commutative diagrams. In *The Spanish Prisoner* (1997), we
see someone leafing through a notebook filled with math, and one of the pages
contains a commutative diagram.

In *Antonia's Line* (1995) the wunderkind of the family, Thérèse, becomes a
math professor. In one scene, she is working on a blackboard containing a
commutative diagram (figure 12.5).

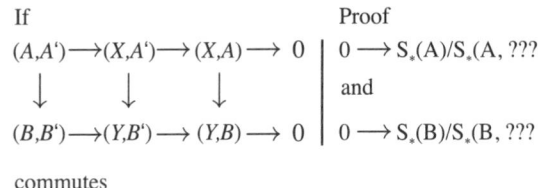

If

$$(A,A') \longrightarrow (X,A') \longrightarrow (X,A) \longrightarrow 0$$
$$\downarrow \qquad \downarrow \qquad \downarrow$$
$$(B,B') \longrightarrow (Y,B') \longrightarrow (Y,B) \longrightarrow 0$$

commutes

Proof

$$0 \longrightarrow S_*(A)/S_*(A, ???$$
and
$$0 \longrightarrow S_*(B)/S_*(B, ???$$

Fig. 12.5 The mathematical prodigy Thérèse, in *Antonia's Line*.

1:11

*THÉRÈSE: We can assume that the singular chain complex of the empty set
equals zero. With theorem 5.8 this implies that the nth homology group is the
same as the nth relative homology group, if we take as the subspace the empty
set. Now we can construct a functor from the category of Top² to the category
of chain complexes. Define the functor S* as follows: S* of the ordered pair
(X, A) becomes the quotient of the singular chain complex of the space X by
the complex of A.*

This seems close to making mathematical sense. But only close.

Part II
Mathematics

Watch those clips!

In this second part of the book, we incorporate the material from many movies into mathematical surveys. It is not essential to have seen the clips beforehand. On the other hand, it definitely wouldn't hurt. Fortunately, many of the clips are available online from the usual sources, and we maintain a list of links to these clips on our website.

Chapter 13
Beautiful Math, or Better Off Dead

There are two kinds of people: those who think that math is beautiful; and those who think that those who think that math is beautiful are crazy. On occasion, one of the former meets one of the latter, and chances are that they'll talk about mathematics. These conversations tend to be short. For example, in *Teresa's Tattoo* (1994), Teresa, a math PhD, is attending a faculty party, and Rick looks to flirt with her:

0:11
RICK: You know, Bruno tells me that you're in the Physics Department?
TERESA: Math.
RICK: Math Department ...
TERESA: See ya.
RICK: Bye.

Sometimes, however, the math fan will be intent on sharing her passion. The level of success can vary dramatically. We consider here three approaches, as exemplified in the movies.

13.1 The Direct Approach

A true math fan always dreams that the uninitiated will be receptive. So they're forever tempted to try the direct approach, even if it seldom seems to work. The fantasized encounter is presumably something akin to the hilarious classroom scene in *Better Off Dead* (1985). Here, we can admire Mr. Kerber in full flight (figure 13.1).

Fig. 13.1 Mr. Kerber, lecturing to his adoring fans.

0:22

MR. KERBER: The three cardinal, trapezoidal formations, hereto made orientable in our diagram by connecting the various points, HIGK, PEGQ and LMNO, creating our geometric configurations, which have no properties, but with location.

CLASS: Aah!

MR. KERBER: ... are equal to the described triangle CAB quintuplicated. Therefore, it is also the five triangles composing the aforementioned NIGH each are equal to the triangle CAB in this geometric concept [laughter]! Therefore, in a like manner, the geometric metaphors can derive a repeated vectoral sum. This was your assignment. I would like to see the results, please. Take them out—Sophia.

CLASS: Oh! Oh! Me!

MR. KERBER: And—Buster.

CLASS: Me! Me! Please!

MR. KERBER: And Beth. And ...

CLASS: Oh, please, please! [The bell rings to end the class.] Aww!

MR. KERBER: Now, now, now. I'll see you all tomorrow. Just remember to memorize pages 39 to 110 for tomorrow's lesson.

ALL: All right!

STUDENT: Cool!

SECOND STUDENT: Mr. Kerber really just makes geometry live!

However, even in a fantasy world, a receptive audience can be pushed too far. Such is clearly the case with Pierre Richard's frenetic lesson in *Lucky Pierre* (1974) (figure 13.2).

Fig. 13.2 Lucky Pierre and his lucky students.

And there's not much to say about Eddie Murphy's efforts to teach unified field theory to the eight-year-old kids in *Meet Dave* (2008) (figure 13.3).

Fig. 13.3 Getting a youthful start on unified field theory.

Of course, the reality is much likely to resemble Professor Larkin's experience in *The Mirror Has Two Faces* (1996); see chapter 11. Similarly, Mikey doesn't fare very well with Sandy, as they discuss their homework in *The Ice Storm* (1997):

0:24
SANDY: Hey, Mikey!
MIKEY: Yeah?
SANDY: Geometry?

MIKEY: Sure, anything but this English.
SANDY: How come you're so good at math but not at English?
MIKEY: I'm not good at math. Just good at geometry. It's like, you know when they say "two squared"? You think it means 2 times 2 equals 4? But they really mean a square. It's really space, it's not numbers, it's space. And it's perfect space, but only in your head, because you can't draw a perfect square in the material world. But in your mind you can have perfect space. You know?
SANDY: Yeah. But I just need some help with my homework.

13.2 The Poetic Approach

As suggested by the above, the direct approach tends to fall flat with the uninitiated. However, more subtly hinting at the hidden beauty of mathematics can work very well. Here are five scenes which exhibit what we call the poetic approach.

In our first scene, Gerald Lambeau, the Fields-medal-winning mathematician in *Good Will Hunting* (1997), is trying to win over a female student:

0:29
LAMBEAU: A difficult theorem can be like a—symphony. It's very erotic.
STUDENT: Wow.

Well, perhaps that "Wow" was less than sincere. Mathematician Paul Rivers (Sean Penn) is more successful with Cristina (Naomi Watts) in *21 Grams* (2003):

1:11
PAUL: There is a number hidden in every act of life, in every aspect of the Universe. Fractals, matter—that there's a number screaming to tell us something. Am I boring you?
CRISTINA: No, no.
PAUL: I'm sorry. I guess I try to tell them that numbers are a door to understanding a mystery that's bigger than us. How two people, strangers, come to meet. There's a poem by a Venezuelan writer that begins: "The Earth turned to bring us closer. It turned on itself and in us until it finally brought us together in this dream."
CRISTINA: That's beautiful.
PAUL: There's so many things that have to happen for two people to meet and—Anyway, that's—that's what mathematics are.

The poetic approach indeed works very well for Paul: he gets the girl.

In *Enigma* (2001), Tom Jericho is a brilliant mathematician (modeled on Alan Turing, one of the top codebreakers at Bletchley Park in the United Kingdom during WWII). He poeticizes to his girlfriend Claire about mathematics:

0:21

CLAIRE: Why are you a mathematician? Do you like sums?

TOM: I like numbers. Because with numbers, truth and beauty are the same thing. You know you're getting somewhere when the equations start looking beautiful and you know the numbers are taking you closer to the secret of how things are.

Well, it turns out that Claire is a foreign spy, after Tom's secrets. Still, she seemed to appreciate the speech.

Our next scene is from *Smilla's Sense of Snow* (1997). Smilla, the Greenlander ice-maiden heroine, is sharing a glass of wine with Andreas:

0:48

SMILLA: The only thing that makes me truly happy is mathematics, snow, ice, numbers. To me the number system is like human life. First you have the natural numbers, the ones that are whole and positive like the numbers of a small child. But human consciousness expands and the child discovers longing.

Do you know the mathematical expression for longing? Negative numbers, the formalization of the feeling that you are missing something. Then the child discovers the in between spaces, between stones, between people, between numbers and that produces fractions, but it's like a kind of madness, because it does not even stop there, it never stops. There are numbers that we can't even begin to comprehend. Mathematics is a vast open landscape. You head toward the horizon and it's always receding, just like Greenland.

Smilla makes a great speech, and it seems to have the desired effect upon Andreas. True, Andreas was already totally in love with her, but let's give a little credit to Smilla's mathematical musings.

Finally, in *A Pure Formality* (1994), novelist and murder suspect Onoff (Gérard Depardieu) is interviewed by a police inspector (Roman Polanski). Onoff muses on the ideals of mathematics:

1:15

INSPECTOR: Ever seen this man, Mr. February?

ONOFF: Seems to me I know him.

INSPECTOR: Who is he?

ONOFF: My math teacher when I was in high school.

INSPECTOR: What else?

ONOFF: It's to him I owe my passion for numbers, for symmetry, for geometric reasoning. Professor Trivarchi: he set forth theorems as if they were fables. His words didn't go through our ears, they went directly to our minds. And I was truly moved: "Two parallel lines can never meet. Nonetheless, it is possible to imagine the existence of a point so far out in space, so far into infinity that we can believe and acknowledge that our two lines do in fact meet there. We shall call this point the ideal point."

The inspector is a fan of Onoff's, and is very impressed by his speech; not so impressed, however, that he doesn't later charge Onoff with murder. Well, he may be a murderer, but Onoff gives a beautiful presentation of the subtle idea behind *projective geometry*. We hope he'll get time off for good (mathematical) behavior.

13.3 The All-Singing, All-Dancing Approach

For those who find poetry too subtle, an alternative is to flamboyantly break into a song and dance. It may not be a method employed often to sell mathematics, but when it is the results can be unforgettable.

In *Are You With It?* (1948), Donald O'Connor plays Milton Haskins, an actuary who misplaces a decimal point, resigns in disgrace, and hooks up with a carny. After a drink, Milton explains that dancing is "merely a question of applied mathematics." He demonstrates $3 \times 2 = 6$, then $6 \times 2 = 12$, followed by "the square root of that raised to the fourth power." Milton then launches into a terrific tap dancing routine. (He appropriately refers to a momentary stumble as "calculus.")

In *Merry Andrew* (1958), Danny Kaye plays a teacher seeking to enthuse his students. His leading them in a song about Pythagoras's theorem seems to do the trick (figure 13.4); see chapter 14.

Fig. 13.4 A merry Danny Kaye leads his young charges in a Pythagorean singalong.

Merry Andrew is difficult to top, but *30 Virgins and Pythagoras* (1977) may just do it. This remarkable movie stars Czech heartthrob Karel Gott as a singing math teacher, and his classes are certainly like no other. As in *Merry Andrew*, Pythagoras's theorem seems to be the theorem of choice to sing about (figure 13.5).

Fig. 13.5 Karel Gott, teaching a very tuneful Pythagoras.

13.4 Any Place, Any Time

Whatever the approach, a math fan tends to choose the moment carefully. However, it is not strictly necessary. In *Stand-In* (1937), mathematician Atterbury Dodd (Leslie Howard) extols mathematics in masterly fashion, while engaging in accountancy:

2:55
ATTERBURY [dictating]*: I am honored by your invitation to address my alma mater. In the past, however, I have encountered difficulties persuading my hearers that the science of mathematics is not one of mere figures and geometrical designs.*
CLERK: 3 o'clock, Mr. Dodd.

ATTERBURY: One minute past, you are late—but a science more important to life than food and drink, a science—[to another clerk] *that account should have been out last night—without which there could be no music, no poetry, no art.*

CLERK [showing Dodd a balance]: *This balance, Mr. Dodd.*

ATTERBURY [immediately]: *There is an error in the addition. The total should be 1296321—the flight of the bird.*

CLERK: The adding machine ...

ATTERBURY: Have the adding machine fixed—the leap of the salmon, the rhythm of the dance, all are mathematical.

Chapter 14
Pythagoras and Fermat at the Movies

Pythagoras's theorem and *Fermat's last theorem* are two of the superstars of mathematics. In this chapter we'll see what the movies can, and cannot, teach us about these megatheorems.

14.1 Pythagoras's Theorem

In *The Wizard of Oz* (1939), the Scarecrow is off to see the Wizard, hoping to get a brain. The Wizard grants the Scarecrow his wish, more or less, by awarding him a Doctor of Thinkology. The Scarecrow then tests his newly gained intelligence by having a go at some mathematics:[1]

1:26
THE SCARECROW: The sum of the square roots of any two sides of an isosceles triangle is equal to the square root of the remaining side—Oh, joy! Rapture! I've got a brain!

The Scarecrow is definitely impressed, but nobody actually familiar with Pythagoras's theorem would be.[2] Let's state clearly the Scarecrow's errors: first it should be "squares" instead of "square roots"; second, "any two sides" should be "the shorter two sides"; third, "isosceles triangle" should be "right-angled triangle."[3]

---

[1] In the original children's story by L. Frank Baum, the Scarecrow wisely refrains from testing out his new brains (made of bran).

[2] Actually, the Wizard in Baum's story is similarly unimpressed, as clearly is Baum himself, with the whole concept of granting brains to someone. The chapter in which the Scarecrow is brained is titled "The Magic Art of the Great Humbug."

[3] In *The Simpsons* episode "$pringfield (or, How I Learned to Stop Worrying and Love Legalized Gambling)" (1993), Homer finds some Henry Kissinger glasses. Deciding that he is now smart, he promptly recites the Scarecrow's Pythagoras line. To his surprise, he receives a response:

NEARBY MAN: That's a right triangle, you idiot!
HOMER: D'oh!

The Scarecrow's errors removed, we now have:

Pythagoras's Theorem (Non-Scarecrow Version)
The square of the hypotenuse of a right-angled triangle
is equal to
the sum of the squares of the two shorter sides.

Of course, for many people Pythagoras's theorem amounts to the mantra: "*a* squared plus *b* squared equals *c* squared." In symbols,

$$a^2 + b^2 = c^2 .$$

However, a triangle must be hidden in that equation somewhere. And, actually the "squares" really are (the areas of) squares. These squares abound in the song "Thanks, Mr. Pythagoras," from the remarkable *30 Virgins and Pythagoras* (1977) (figure 14.1).

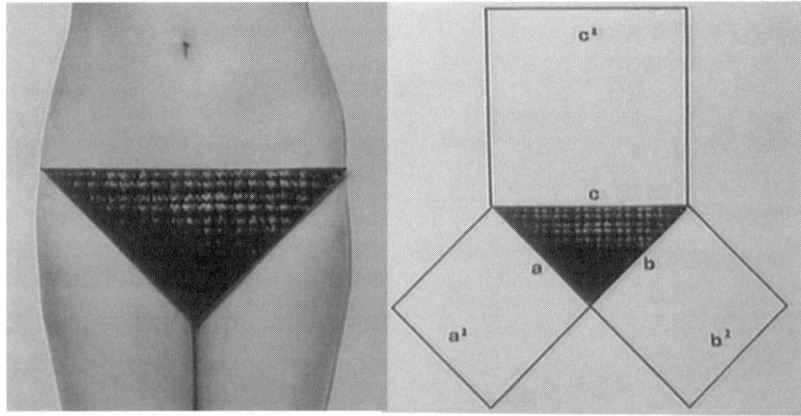

Fig. 14.1 An inspired choice of triangle, and its Pythagorean squares.

In *Merry Andrew* (1958), Danny Kaye plays the new teacher at a boarding school, where the boys have been raised on rote learning. He seeks to excite them, and succeeds brilliantly by singing and dancing "The Square of the Hypotenuse."[4] It is a difficult movie to obtain, but the scene is a must-see.

Though scientific laws may change
and decimals can be moved,
the following is constant
and has yet to be disproved.
The square of the hypotenuse
of a right triangle

---

[4] The lyrics of the song are by Johnny Mercer, more famous for *Moon River* and *Jeepers Creepers*.

is equal to the sum of the squares
of the two adjacent sides ...

Andrew finishes the dance by arranging himself, the boys and a string of flags
into a large, very impressive right-angled triangle (figure 14.2).

Fig. 14.2 ... is equal to the sum of the squares of the two adjacent sides.

Andrew is too busy being merry to actually offer the boys a *proof* of
Pythagoras's theorem. In *The Man Without a Face* (1993), a disfigured and
much less merry Mel Gibson is also teaching geometry, and he does provide
the proof of a Euclidean theorem. Alas, Mel's proof is incorrect; see chapter 18
("Money-Back Bloopers"). So, though Mel demands that his student state
Pythagoras's theorem, it's probably a relief he didn't also demand a proof.

In *Late Bloomers* (1996), we hear an actual proof of Pythagoras's theorem, by the teacher, Miss Groshardt.[5] The proof itself is a famous one by way
of similar triangles, and small glimpses of it appear on the blackboard. The
closest to a visual proof is the background diagram in figure 5.1, and a similar
diagram in the episode "Youth + Vampire" of the anime series *Rosario +*

---

[5] Sheesh! The movie is actually an extraordinarily tedious lesbian chick flick. However, it
did introduce us to the unforgettable expression "She's aardvarking with the hypotenuse."

Vampire (2008). In this episode, we see part of the famous windmill proof of Pythagoras's theorem behind the teacher Ririko. Given Ririko's, um, enthusiastic teaching style (figure 14.3), we regret that she didn't work through the proof in detail.

Fig. 14.3 A voluptuous presentation of the Pythagorean windmill.

14.2 Fermat's Last Theorem

Pythagoras's theorem means that every right-angled triangle gives rise to three positive numbers a, b, and c that satisfy the equation

$$a^2 + b^2 = c^2 \, .$$

Pythagorean triples are triples of *natural numbers* that satisfy the equation. The simplest Pythagorean triple, which most people will recall from schooldays, is

$$3^2 + 4^2 = 5^2 \, .$$

Are there others? Yes, lots.[6] And, it was probably the existence of Pythagorean triples that made mathematicians wonder about other powers. Are equations such as

---

[6] Any square multiple of a Pythagorean triple is again a Pythagorean triple. For example, multiplying the previous triple by 2^2, we find $6^2 + 8^2 = 10^2$. Here is a simple trick to derive infinitely many Pythagorean triples that are not so related.

Pick a natural number n and consider the identity $(2n + 1) + n^2 = (n + 1)^2$. This will be a Pythagorean triple as long as the odd number $2n + 1$ is a square. But that's easy to arrange: just pick any odd square, and then choose n accordingly. For example, the odd square $9 = 3^2$ arises from choosing $n = 4$, and then the identity can be written as $3^2 + 4^2 = 5^2$. Similarly, the odd square $25 = 5^2$ (from the choice $n = 12$) leads to the Pythagorean triple $5^2 + 12^2 = 13^2$. In fact, all possible Pythagorean triples can be obtained from similar identities.

$$x^3 + y^3 = z^3, \quad x^4 + y^4 = z^4, \quad x^5 + y^5 = z^5, \quad \ldots$$

also satisfied by some natural number triples? Captain Picard summarizes the story of this problem at the beginning of the *Star Trek* episode "The Royale" (1989):

PICARD: Fermat's last theorem. You familiar with it?
RIKER: Vaguely. I spent too many math classes daydreaming about being on a starship.
PICARD: When Pierre de Fermat died, they found this equation scrawled in the margin of his notes. $x^n + y^n = z^n$ where $n > 2$, which he said has no solution in whole numbers. But he also added this phrase: Remarkable proof.
RIKER: Yeah, it is starting to come back to me. There was no proof included.
PICARD: Hmm, and for 800 years people have been trying to solve it.
RIKER: Including you.
PICARD: I find it stimulating. Also, it puts things in perspective. In our arrogance, we feel we are so advanced and yet we cannot unravel a simple knot tied by a part-time French mathematician working alone without a computer.

Later, Picard sums up his crew's confusing adventure:

PICARD: Like Fermat's theorem, it's a puzzle we may never solve.

Captain Picard has most of the story correct. To begin, we have

---

Fermat's Last Theorem
If n is a natural number greater than 2,
then $x^n + y^n = x^n$ has no natural number solutions x, y, and z.

In 1637 the "part-time" (but extremely gifted) mathematician Pierre de Fermat really did jot down the following note in one of his math books: "I have a truly marvelous proof of this proposition which this margin is too narrow to contain." It is also true that for a long time nobody was able to prove or disprove this claim. However, 357 years later, and five years after the *Star Trek* episode, Princeton University mathematician Andrew Wiles finally managed to come up with a proof.

It's hard to blame the scriptwriter for the above lines of Captain Picard. In 1989 few mathematicians, except for Wiles, held much hope that Fermat's last theorem would soon be solved. It seemed entirely plausible that mathematicians would still be chasing the theorem centuries later.

In any case, the *Star Trek* team catches on fast. In the *Deep Space Nine* episode "Facets" (1995), Dax raises the problem:

DAX: I've been working on finishing your proof of Fermat's last theorem ...
O'BRIEN/TOBIN: You have?
DAX: It's the most original approach to the proof since Wiles's over three hundred years ago ...
O'BRIEN/TOBIN [flattered]: *Thanks ...*

It is not clear whether Dax is implying that Wiles's proof is wrong, or just that she has been working on a new proof. If the former, then maybe Picard was right after all.

Either way, to prove Fermat's last theorem is hellishly difficult. In *Bedazzled* (2000), Elizabeth Hurley plays the Devil, and in one of her incarnations she is leading astray a classroom of boys (figure 14.4):

Fig. 14.4 A devilish homework problem.

1:01

DEVIL: Okay, boys. Tonight's homework. [She walks to the blackboard.] *Algebra: x to the nth plus y to the nth equals z to the nth. Well, you're never going to use that, are you?* [She wipes out the problem, and everybody laughs.]

The "SHOW YOUR WORK" on the board appears to be a very funny jibe at Fermat, as if he were a negligent schoolboy for not including his proof.

In *The Simpsons* episode "Homer^3" (1995), we see the following equation floating in Homer's 3D world:

$$1782^{12} + 1841^{12} = 1922^{12}\,.$$

If the equation were correct then it would be a counterexample, demonstrating Fermat's last theorem to be false. And, checking on a calculator, it may well seem to be correct: *The Simpsons* team have worked hard on that one. However, closer inspection makes clear that the equation cannot possibly be correct: the left side is definitely odd, and the right side is definitely even. In fact, properly evaluating the right and left sides gives

$$2, 541, 210, 258, 614, 589, 176, 288, 669, 958, 142, 428, 526, 657$$

and

$$2, 541, 210, 259, 314, 801, 410, 819, 278, 649, 643, 651, 567, 616 .$$

Two huge numbers, differing by a huge amount, but nonetheless agreeing in the first nine digits.

In a later episode, "The Wizard of Evergreen Terrace" (1998), the Simpsons have another poke at Fermat's last theorem:

$$3987^{12} + 4365^{12} = 4472^{12} .$$

This time, both sides are even numbers, so the previous even-odd trick doesn't work. However, Wiles can rest assured that the equation is nonetheless false: the left side is divisible by 3 (since both 3987 and 4365 are), whereas the right side is not (since 4472 is not).

14.3 Fermat's Last Tango

Fermat's Last Tango (2001) is a mathematical musical, perhaps the first ever.[7] It is the work of husband-and-wife team Joshua Rosenblum and Joanne Sydney Lessner, and in 2000 was presented off Broadway by the York Theatre Company. A DVD of a live performance is available from Boston's Clay Institute.

To do *Fermat's Last Tango* justice, we would have to include the whole transcript. It probably contains more mathematicians and more higher mathematical references than all other movies combined. And it is terrific fun. We'll reluctantly restrict ourselves to summarizing the story, and detailing some of the best dialogue.

The story begins with Daniel Keane (=Andrew Wiles) announcing his proof of *the* theorem. Daniel is an immediate celebrity, with everybody wanting to meet him. In fact, even Pierre de Fermat stops by, claiming to have

. . . discovered a marvelous method of time travel, which this attic is not large enough to . . .

In reality, Fermat is visiting from the AfterMath, where the really famous mathematicians from the past are basking in their glory. Keane has just earned his place in the AfterMath, and ostensibly Fermat has come to invite Keane to meet with Pythagoras, Euclid, Newton, and Gauss. When we first encounter the Fab Four, they are harmonizing:

0:24
Sing me to Symmetry
Muse of the Mathematic
We worship all equations

---

[7] And perhaps not. The International Movie Data Base (IMDB) lists a Brazilian musical, *Matemática Zero, Amor Dez* (1958). We have been unable to determine absolutely anything about this movie, much less obtain a copy.

the simple and quadratic
Algebra, Geometry, Set and Number Theory
All admired equally
In our Purgatory and Pythagorean secret society.

However, it turns out that the preponderance of large egos means that not all is harmonious in the AfterMath. Also, it turns out that these mathematicians, great as they are, are from mathematically simpler times: Keane's proof is too modern and too complicated for them. Moreover, Fermat is not at all happy that somebody else has found a proof of *his* theorem. It all explodes when Fermat gleefully announces that he has discovered a "big, fat hole" in Keane's proof:[8]

0:33
KEANE: A hole?
EUCLID and NEWTON: A hole?
KEANE: My proof contains a hole?
FERMAT: I didn't want to be the one to say.
I know this is upsetting,
please show some self-control,
but your proof contains a big, fat hole!

PYTHAGORAS:	*EUCLID, NEWTON, GAUSS:*
Make no mistake about it,	*ooh, ooh*
if this defect you can mend,	*shoo-bi-doo-doo-wah,*
an offer of admission	
we will cheerfully extend.	*ooh, ooh,*
But until the time arrives	*shabada-boo,*
you are forbidden to enroll,	*ah, ooh, olé.*
'cause your proof contains a big, fat hole!	

A despondent Keane retreats to his attic to try to fix his proof. Fermat hangs around, to gloat and to confuse him.

0:42
FERMAT:
I don't intend to lose my immortality,
and share with some two-bit professor!
A person who in raw intelligence
is so obviously my lesser.

Now we know! And the other residents of the AfterMath know, too; they are getting very annoyed with Fermat, and hope to "wipe his smirk from his face." They decide to update their mathematical knowledge and help Keane fix the hole. In the end they fail, and in admitting defeat they acknowledge

---

[8] There was indeed a hole in Andrew Wiles's proof. It was corrected by Wiles, in collaboration with mathematician Richard Taylor.

Fig. 14.5 Pythagoras, Newton, Gauss, and Euclid, with Fermat on air double bass.

Keane's amazing achievement in getting as far as he did. They admit Keane to the AfterMath.

Luckily, Keane also has a staunch supporter in his wife Anna. Not so lucky for Anna, as she sees herself as a "math widow" (as indeed are all math wives and math husbands). However, just as Keane is about to give up, Anna, who doesn't know a thing about mathematics, remarks:

0:79

ANNA: It's often said, "Within your failures lie the seeds of your success."

This leads Kean to reconsider an approach that he had previously given up on. Now, in a new light, this old approach is the answer to his prayers. And they lived happily ever after.

But what about Fermat and his proof? Did he also have a proof? Well, even he is happy after Keane, with a bow to Fermat, pronounces:

0:83

KEANE: His proof was so simple,
he claimed it was simple,
and I believe that what he said was true.
But since through the centuries,
no one has conjured it,

you could almost say Fermat was
the most brilliant mathematician
the world has ever known.

Fermat's Last Tango is a must-see. It is historically accurate, it doesn't dodge the mathematics, and it is great fun. For a detailed review from a mathematician's perspective, see Robert Osserman's article.[9] See also the musical's webpage, hosted by the Clay Institute.

The original title of the musical was *Proof*. It was changed after David Auburn's play of the same title, about a mathematics professor and his daughter, became a hit. The following remarks are by Lessner:[10]

"When we started working on *Fermat's Last Tango* in December 1996, we had no idea that it would eventually be perceived as part of an unprecedented trend of stage works about math and science. We were tremendously excited to see, last season, the success of *Copenhagen* and *Proof*,[11] among others, and we, along with [York artistic director] Jim Morgan realized that the time for a musical that takes math as its milieu was now or never.

"It's about obsession, a real-life quest that lasted 30 years, the fruit of which yielded the single most extraordinary contribution to modern mathematics. It's *Rocky, Don Quixote*, even *The Fantasticks*—boy gets proof, boy loses proof, boy gets proof. We knew going in that the subject matter could be potentially forbidding, so we focused on keeping the piece accessible and, above all, fun.

"Josh had long toyed with the idea of writing a 'catchy' opera, and it turned out that the tuneful and rhythmic elements of our show superseded the fact that it was through-sung and had some classical influences. By the middle of the process it became clear that *Fermat's Last Tango* was a musical, not an opera; in addition to the tango, there are several recognizable dance forms. There's a rag, a grand waltz, and even a hoedown for the four luminaries [*Pythagoras, Euclid, Newton, and Gauss*] who populate the AfterMath. I like to call them the 'singing and dancing dead mathematicians.' Our director, Mel Marvin, likens them to characters out of *Monty Python's Flying Circus*—so serious and full of themselves that they are incongruously funny."

---

[9] Robert Osserman, Fermat's last tango, *Notices of the American Mathematical Society* 48 (2001), 1330–1332.

[10] Kenneth Jones, From page to stage: Einstein's dreams, the Musical, gets NYC reading, *Playbill On Line*, January 16, 2003.

[11] In 2005, an excellent film version of *Proof* was released, starring Gwyneth Paltrow and Anthony Hopkins. *Copenhagen* was released as a film in 2002.

Chapter 15
Survival in the Fourth Dimension

In *The Simpsons* episode "Homer³," Homer is seeking to hide from Marge's sisters and discovers a portal to the third dimension. Rather than face the sisters, he takes his chances in the mystery world, and gets trapped.

LISA: So where's my dad?
PROFESSOR FRINK: Well, it should be obvious to the most dimwitted individual who holds an advanced degree in hyperbolic topology, m'hey, that Homer Simpson has stumbled into [lights go off] *the third dimension!*
LISA [switches the light back on]: *Sorry.*
FRINK: Here is an ordinary square.

Fig. 15.1 Explaining the Frinkahedron.

CHIEF WIGGUM: Whoa, whoa. Slow down, egghead.
FRINK: But suppose we extend the square beyond the two dimensions of our universe along the hypothetical z-axis, there.
EVERYONE: Gasp!
FRINK: This forms a three-dimensional object known as a "cube," or a "Frinkahedron," named in honor of its discoverer. M'hey, m'hey.
HOMER: Help me! Are you helping me, or are you just going on and on?
FRINK: Oh, right. And, of course, within we find the doomed individual.

Looking for a footnote? There isn't one. "Homer³" is actually the title of the episode.

Of course, the two-dimensional Homer is totally unprepared to meet the challenges of a three-dimensional world.[1] Similarly, we suspect that most of our three-dimensional readers would be very lost if suddenly thrown into a four-dimensional world. With this in mind, and to alleviate the great concern created by this ever-present danger, we offer here our survival guide to the fourth dimension.

15.1 Time, Space, Both, or What?

In the *Devil Girl From Mars* (1954), the devil girl Nyah is describing the ultimate weapon:

0:28
NYAH: As fast as matter was created, it was changed by its molecular structure into the next dimension, and so destroyed itself.
PROFESSOR HENNESSEY: So there is a fourth dimension!

The Professor seemed very convinced by Nyah, though we're not sure why. Indeed, ever since Einstein, "the fourth dimension" has been a hugely popular topic of discussion, but seldom accompanied by much understanding. As part of Einstein's theory of relativity, the three spatial dimensions are woven together with time to create a four-dimensional space-time universe, a beautiful and mysterious concept. It's also not at all what we wish to discuss.

For mathematicians, the fourth dimension is (usually) different, and it was a mathematical reality at least fifty years prior to Einstein's space-time ingenuity. To describe it, we first note that the position of a point in everyday three-dimensional space (our three-dimensional Euclidean space) is determined by its three coordinates (x, y, z). Analogously, a point in four-dimensional space is pinned down by four coordinates (x, y, z, w). And if you want more dimensions, it's easy: for example, a point in seven-dimensional space is simply pinned down by seven coordinates (x, y, z, w, p, q, r).

Now, what all these dimensions *mean* is a matter of context and of debate. However, the *mathematics* of higher dimensions is quite easy. These mathematical worlds are not fundamentally different from our three-dimensional world, and they can be played with and analyzed in very similar and familiar ways. Our goal is to describe how, using movie math as our platform.

What then has all this to do with time being the fourth dimension? Not a whole lot. In relativity, there are also four dimensions, pinned down by the four coordinates (x, y, z, t). However, whereas in our four-dimensional world all the dimensions are alike, this is definitely not the case for space-time: the time dimension is fundamentally different from the space dimensions. For example, we can easily move back and forth in space, whereas moving back in time requires the purchase of a time machine. In the following, we'll

---

[1] All the scenes in the three-dimensional world can be experienced in full 3D, as part of the IMAX movie *Cyberworld* (2000).

focus on a fourth spatial dimension, steering clear of the special and trickier concept of time as the fourth dimension.[2]

15.2 The Hypercube Via Analogy

The movie *Cube 2: Hypercube* (2002) is the sequel to *Cube* (1997); see chapter 6 ("Escape from the Cube"). As in *Cube*, some lucky people wake up in a very hostile world. This world consists of cubical rooms, interconnected by doors in the walls, ceilings, and floors (figure 15.2).

As the people stumble from room to room, very strange things happen: rooms loop in on themselves; rooms that were adjacent cease to be so; time (there we go!) moves differently in different rooms; and gravity acts in different directions in different rooms. Amidst it all, the lucky people have to be on the lookout for deadly traps.

The main characters are psychologist Kate Filmore, private investigator Simon Grady, computer game programmer Max Reisler, engineer Jerry Whitehall, and the old, senile, and incredibly annoying mathematician, Mrs. Paley.

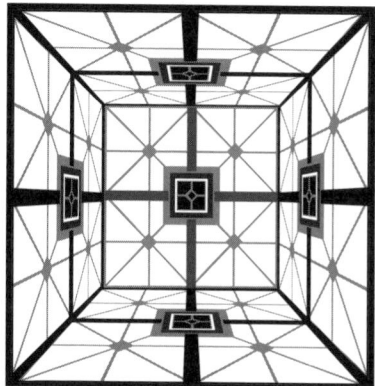

Fig. 15.2 One of the cubical rooms in *Cube 2*.

0:25
MRS. PALEY: It's a tesseract.
JERRY: Christ, she's losing it.

---

[2] There is obviously no shortage of movies with plots tied to time travel and the like, but there is seldom much explicitly mathematical in the telling. One beautiful movie that gives the sense of time and its irreversibility, as well as containing some clever geometric pondering, is the aptly named *The 4th Dimension* (2006). There is also the terrific *Futurama* episode, "The Late Philip J. Fry" (2010), in which Professor Farnsworth, Fry, and Bender accidentally go forward 10,000 years into the future. The professor's time machine cannot go backward in time, and so they are stuck there. They decide to go forward instead, and discover that time loops around, so they arrive back safely (eventually, after a second trip through the loop).

MRS. PALEY: Isn't it beautiful?
KATE: Isn't what beautiful, Mrs. Paley? ·
[Mrs. Paley points at a diagram on the floor.]
JERRY: Holy Shit! If you look at it from just the right angle—What did you call it again Mrs. Paley?
MRS. PALEY: It's a tesseract, sweetheart.
JERRY: Tesseract?—Tesseract. It's a tesseract!—A tesseract, it's another name for a hypercube—a four-dimensional cube. All the elements are there. I mean, rooms repeating, rooms folding in on themselves. Teleportation. It could all very well add up.

   Look. Here. See? [He begins a Frink-style explanation.] *Let's call one dimension length and represent that with a simple line. Now, two dimensions are length and width, which is represented by a simple square. If we extend this by one more dimension we get a cube, which has three dimensions: length, width, and depth.* [Figure 15.3]

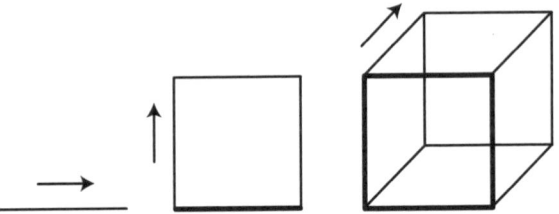

Fig. 15.3 Growing a cube.

JERRY: Here's the really funky part. If you take this cube and we extend it one more dimension, we get ...
KATE: A tesseract.
MAX: I thought time was considered to be the fourth dimension. [No, no, no!]
JERRY: Sure that is one idea. But what if you have a fourth spatial dimension?

Jerry's Frink-style lecture is a very accurate description of how to construct a four-dimensional cube, what mathematicians refer to as a *tesseract* or *hypercube*. Jerry uses the familiar counterparts in three dimensions to motivate the four-dimensional analogy. Such analogizing is one of the most powerful tools for creating and understanding four-dimensional objects.

   Furthermore, the process can be continued to construct a five-dimensional cube, and then a six-dimensional cube, and so on. Also, since a line segment and a square are the first two objects in the process of growing cubes, mathematicians sometimes refer to a line segment as a "one-dimensional cube" and to a square as a "two-dimensional cube." In fact, at times they even refer to a "zero-dimensional cube," which is just a point.

15.3 Picturing the Hypercube

As an application of Jerry's interdimensional analogy, let's see how to visualize higher-dimensional cubes. The first two diagrams of figure 15.3 really are one-dimensional and two-dimensional cubes. However, the third diagram is not actually a three-dimensional cube (unless you've purchased the special hologram edition of this book). Rather, it is a *picture* of such a cube. To make this picture, we first draw a square and then a copy of the square, translated a little in both the horizontal and vertical directions. Corresponding corners of the squares are then connected (figure 15.4).

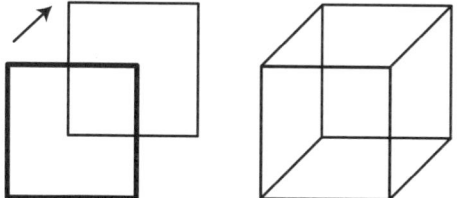

Fig. 15.4 Growing a cube by doubling a square.

This is the same doubling process that produces a square from a line segment, and a line from a single point, suggesting that we can use the same technique to draw a tesseract. Take a picture of the cube and make a new, translated copy. Then connect corresponding corners of the cubes by edges. This results in the picture of a tesseract; see figure 15.5. Continuing in this manner, we can produce flat pictures of cubes of any dimension.

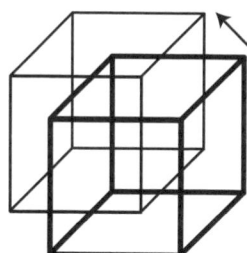 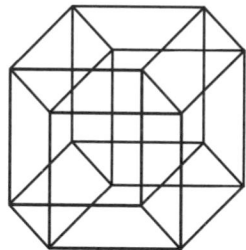

Fig. 15.5 Growing a hypercube by doubling a cube.

The word "HYPERCUBE" is grown in the opening titles of the movie *Cube 2*; see figure 15.6. Very nicely, the method by which the word grows is based upon the above method for growing cubes. The creation of this title sequence is discussed in the DVD extra, "Making of *Cube 2: Hypercube*."

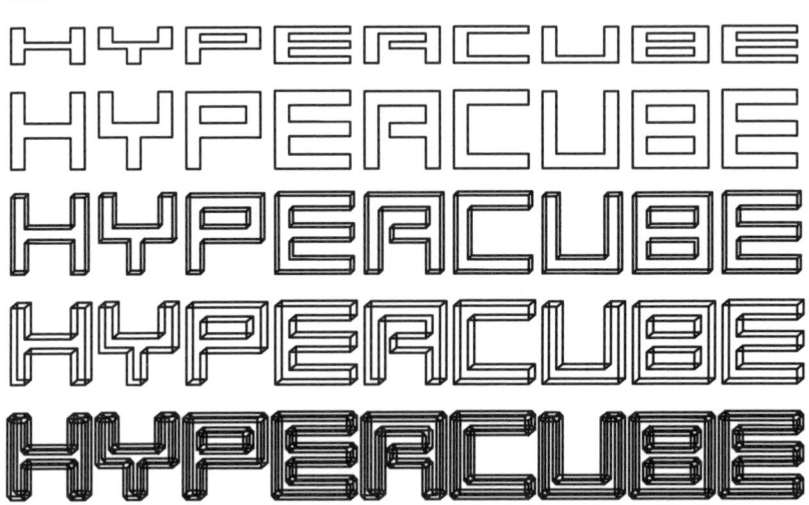

Fig. 15.6 Growing the title of the movie.

Jerry guesses that he and the others are trapped in a four-dimensional cube. He could easily have saved them a lot of trouble, if he had gone further and deduced some of the basic properties of the hypercube.

How Many Rooms?

A one-dimensional cube is bounded by two zero-dimensional cubes: without the jargon, a line segment has two endpoints. Next, a two-dimensional cube is bounded by four one-dimensional cubes (that is, a square has four sides). Then, a three-dimensional cube is bounded by six two-dimensional cubes (that is, a cube has six square faces). If we trust in patterns, this suggests that a four-dimensional cube should be bounded by eight regular cubes. That is indeed the case. We highlight the eight cubical rooms in figure 15.7.

This also suggests that the hypercube world in which Jerry and the others are trapped consists of only eight rooms. These eight cubical rooms are stuck together face to face in the fourth dimension, to form the "surface" of a four-dimensional cube.

This is an important and subtle point! When we speak about a "cube," we may mean a solid cube, for example, a die. However "cube" may also refer to a hollow cube, such as a box. The solid cube is genuinely three-dimensional, whereas the hollow cube is a two-dimensional surface within three-dimensional space.

When Jerry speaks of a hypercube, he actually means a hollow rather than a solid hypercube. This hollow hypercube is actually three-dimensional, though it is the "surface" of a solid, genuinely four-dimensional hypercube.

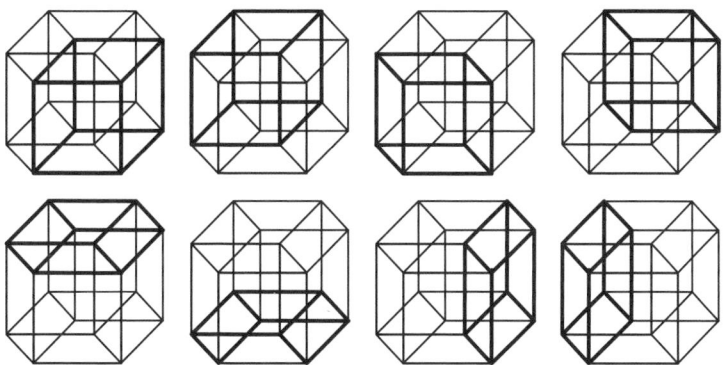

Fig. 15.7 A hypercube is bounded by eight cubes.

This may well boggle the mind, but just relax and trust in what the analogies suggest.

Our picture of the hypercube also shows its corners and edges, and we can see it has 16 corners. Of course, we could easily have guessed this, from the doubling process, and since the lower-dimensional cubes have 1, 2, 4, and 8 corners.

0:29
They find a number on the ceiling.

JERRY: 60,659 rooms, Christ.
KATE: This place must be huge.
MRS. PALEY: Oh yes, yes—in a hypercube, there could be 60 million rooms.
JERRY: She could be right.

Jerry really should know better.

No Way Out!

0:18
JERRY: There's got to be some kind of logic to it. These rooms just seem to repeat. You go in one direction the room just loops back on itself.

Again, by analogy, it is possible to figure out why these cubical worlds should loop in on themselves. For the moment, imagine the victims are one-dimensional beings caught in the edge of a hollow two-dimensional cube. That is, our victims are trapped in the perimeter of a square.

The perimeter of a square is a loop. So, moving in one direction, the one-dimensional beings will pass through the four "rooms" and loop back to where they began. Exactly the same is true for two-dimensional creatures trapped in the surface of a cube; walking in one direction, the creatures will return to their starting square after having passed through three other squares (figure 15.8).

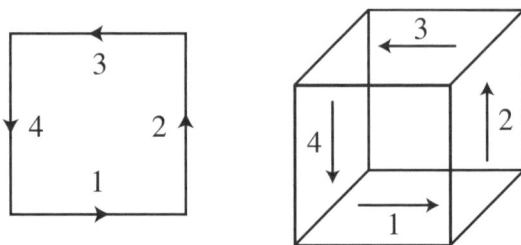

Fig. 15.8 Moving in one direction through four rooms gets you back to where you started.

This looping occurs in two- and three-dimensional cubes, and so by analogy it should be a feature of hypercubes as well. And it is. Moving in one direction, Jerry and the others will return to where they began after traversing a total of four rooms. Such a round trip is highlighted in figure 15.9.

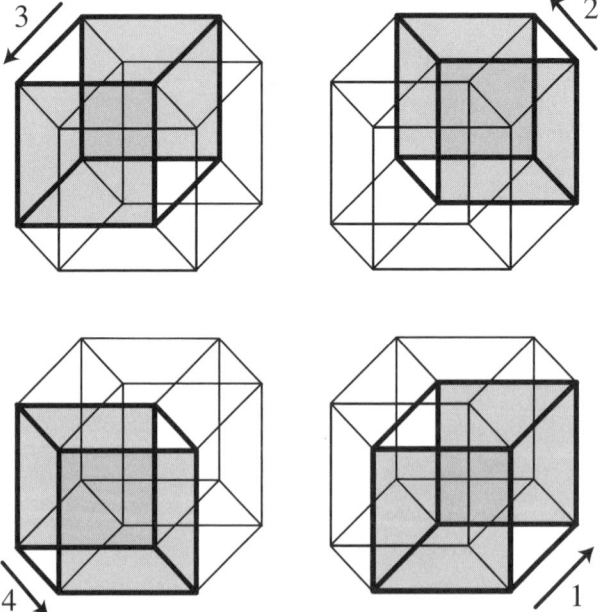

Fig. 15.9 A loop of four cubes, joined along the highlighted squares.

Finally, notice that a hollow cube has no trapdoors offering escape to an outside world. Leaving one room always just leads to winding up in another. So Jerry and the others really are in trouble.

The Schlegel Diagram

Figure 15.10 shows the diagram of the tesseract recognized by the annoying (but knowledgeable) Mrs. Paley. It is clearly a very different picture of the hypercube, and arises from a different inter-dimensional analogy.

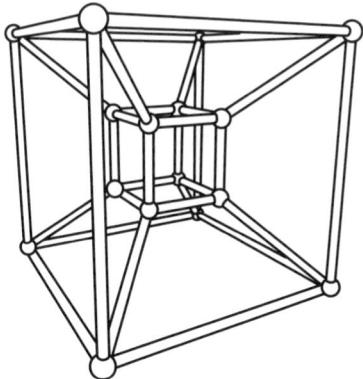

Fig. 15.10 Mrs. Paley's tesseract.

Take a wire frame model of a three-dimensional cube and project it onto a screen. In the resulting shadow, all the corners and edges of the cube are visible. However, some of the faces are somewhat distorted and no longer appear as squares.

Figure 15.11 shows a particularly nice shadow of a cube, known as a *Schlegel projection* of the cube. There is apparently a face missing, but the outer square corresponds to this missing face.

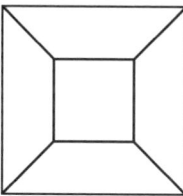

Fig. 15.11 A Schlegel projection of the cube.

We may be lacking a four-dimensional lightbulb, but mathematically there is no difficulty in casting a similar shadow of a wire frame hypercube onto a "three-dimensional screen." This shadow will be a three-dimensional wire frame, a Schlegel projection of the hypercube. Figure 15.10 is a two-dimensional picture of this Schlegel projection, which is what Mrs. Paley discovered. Looked at carefully, the Schlegel diagram exhibits seven distorted rooms. The eighth room has to be represented there somehow, and it is: it corresponds to the outside cube.

This shadow of the hypercube makes a charming reappearance later on in the movie. It is one of the incarnations of the so-called razor sphere, a booby trap that shreds poor Jerry to pieces.

The hypercube mathematics above explains some of the strange phenomena experienced by the group. However, the mathematical aspects are only marginally developed in the movie, and distracting bits and pieces commonly associated with the fourth dimension get mixed into the story. For example, time seems to pass at different speeds in different rooms. Much of the pseudo-science that was supposed to explain what else was going on in the Hypercube is described in the DVD extra, "Making of *Cube 2: Hypercube*."

There is a final piece of very nice mathematics in *Cube 2*. Toward the end, all the rooms collapse onto a single room, of which the opposite sides are identified. So leaving the room through one of the doors results in reentering it through the door in the opposite wall (figure 15.12).[3] This new, strange place is also a famous three-dimensional world, known as the *3-torus*.

Fig. 15.12 The last two survivors of *Cube 2*, in the collapsed hypercube (3-torus).

One of the nicest animations of growing higher-dimensional cubes appears in *Supernova* (2004). A point is expanded into a line, then the line into a square, then the square into a cube, then the cube into the Schlegel projection of a hypercube (figure 15.13), then … oh well … The animations do cheat a little bit with the higher-dimensional pictures, but it is still pretty cool:

---

[3] Similarly stylish looping occurs in *Graveyard Disturbance* (1987), *Pleasantville* (1998), *The Avengers* (1998), *Matrix Revolutions* (2003), and the *Futurama* episode "Bender's Game" (2008).

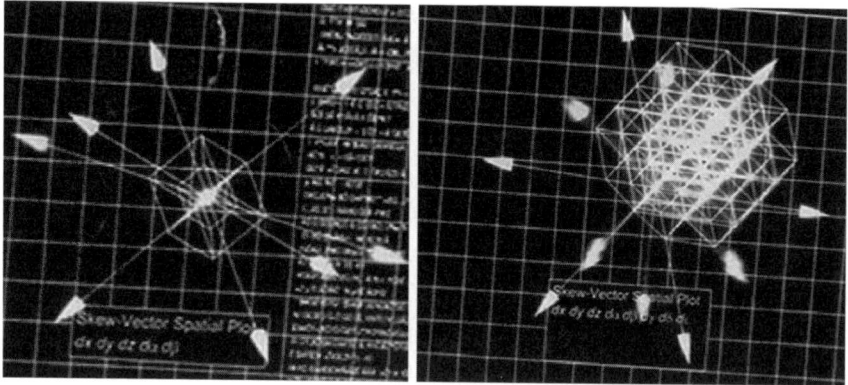

Fig. 15.13 Constructing the picture of a higher-dimensional cube.

15.4 Dimension Drive

The movie *Supernova* also features a "dimension drive," enabling a spaceship to cover vast distances in no time. Does this make any sense? It does! Well, at least mathematically.

Imagine you're a two-dimensional being and trapped in a spherical world, and you want to travel from the North Pole to the South Pole. You are restricted to traveling within your two-dimensional world, along the spherical surface. However, if you have purchased a Dimension Drive (patent pending), you can travel through the surrounding three-dimensional space, taking the considerably shorter straight-line route.[4]

Analogously, imagine that the three-dimensional universe we live in is situated inside a higher-dimensional universe. Then two points that are light-years apart within our world may actually be only centimeters apart when one travels through the higher-dimensional space. If so, a dimension drive would permit faster-than-light travel.

A similar mode of zippy transport is featured in *Event Horizon* (1997).[5] Dr. Weir (Sam Neill) attempts to explain it to the members of a rescue mission:

14:30
DR WEIR: The Event Horizon was the culmination of a secret government project to create a spacecraft capable of faster-than-light flight.
SMITH: Uh, excuse me. See, you can't actually do that.
STARK: The law of relativity prohibits faster-than-light travel.
DR WEIR: Relativity, yes. We can't break the law of relativity, but we can go around it. The ship doesn't really go faster than light. What it does is it

---

[4] Jack, the precocious schoolboy in *The 4th Dimension* (2006), tries this argument out on his teacher. He explains it very well, but she doesn't seem at all impressed.

[5] A similar explanation can also be found in *Déjà Vu* (2006).

creates a dimensional gateway that allows it to jump instantaneously from one point of the Universe to another light-years away.
STARK: How?
DR WEIR: Well, that's difficult to—it's all math.
MILLER: Try us, Doctor.
DR WEIR: Right. Well, uh, using layman's terms, we use a retaining magnetic field to focus a narrow beam of gravitons. These, in turn, fold spacetime, consistent with Weil tensor dynamics until the space-time curvature becomes infinitely large and you produce a singularity. Now, the singularity
. . .
MILLER: Layman's terms.
COOPER: Well, fuck layman's terms. Do you speak English?
DR WEIR: Imagine for a minute that this piece of paper . . . represents spacetime, and you want to get from point A here to point B there. [He marks two points on the paper by punching holes.] *Now, what's the shortest distance between two points?*
JUSTIN: A straight line.

Fig. 15.14 By folding a piece of paper any two points can be made to coincide.

DR WEIR: Wrong. The shortest distance between two points is zero, and that's what the gateway does. It folds space so that point A and point B coexist in the same space and time. [He folds the piece of paper such that the two holes in the paper are superimposed (figure 15.14).] *When the spacecraft passes through the gateway, space returns to normal.* [He passes a pencil through the holes and unfolds the piece of paper.]

Using a dimension drive would allow a spaceship to vanish at one point of the Universe and to reappear at another: definitely handy for avoiding that peak-hour traffic. In fact, a great many miracles, such as having objects vanish and unlinking solid rings, can easily be explained (at least in movie worlds) using higher dimensions. For example, the razor sphere in *Cube 2* makes its appearance in this manner. However, this is a miracle Jerry could have done without.

15.5 Intersections

In *Shrieker* (1997), we meet some monsters that turn out to be intersections of higher-dimensional beings with our three-dimensional world. The hero is Clark, a math student interested in higher dimensions:

0:38
ZAK: So, what are you working on?
CLARK: This? Oh, actually it's multidimensional topography.
ZAK: Please don't ...

Zak is already lost, or bored, or both. (Perhaps Clark describing her work as "topography" rather than *topology* has confused him.) It's a shame, as Clark was all ready to describe the intersection of a four-dimensional object with our three-dimensional world. This information would have made Zak much more useful when the monsters attacked.

Imagine that your friend is a two-dimensional being, living in a plane, and you want to give him a sense of what a cube is. You can show him the cube a slice at a time, by passing the cube through his plane; see figure 15.15.

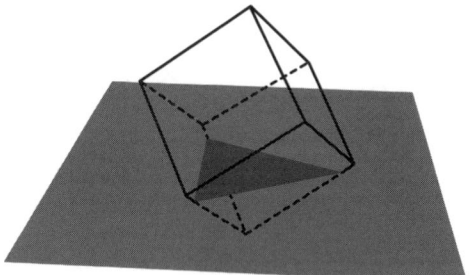

Fig. 15.15 A cube passes corner-first through a plane.

Suppose the cube is lowered down corner first. When this corner first touches the plane, a point is visible, and then an equilateral triangle grows. As the triangle grows further, its corners become chopped off, until eventually the plane contains a perfect regular hexagon. Then everything is repeated in reverse, until the cube disappears (figure 15.16).

Fig. 15.16 The intersections of a cube as it passes through a plane.

Now imagine a hypercube passing through our three-dimensional world, corner-first. Initially we see a point, and then a growing tetrahedron. As the tetrahedron grows further, its corners become chopped off, until eventually we see an octahedron. Then everything is repeated in reverse, until the hypercube disappears (figure 15.17).

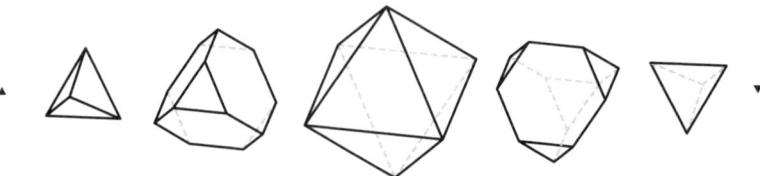

Fig. 15.17 A hypercube passes corner first through our world.

Returning to *Shrieker*, Clark explains her deciphering of some mysterious symbols:

0:39

CLARK: Okay, I've been looking over your notes and you're completely wrong.

ROBERT: Well, that's progress at least. Why am I wrong?

CLARK: You've been treating these symbols as if they were mathematical terms, but they're not.

ROBERT: Then what are they?

CLARK: They're cross-sections. See, the even symbols are two-dimensional cross-sections of a three-dimensional shape. The odd symbols find the angle of the cross-section . . .

ROBERT: This is what I think. There are higher dimensions and things live in them. Sometimes there are weak spots between the higher ones and our own where they can break through. And I think sometimes these breakthroughs must happen spontaneously, not often, but there have been other cases.

CLARK: Like lightning?

ROBERT: That's right.

Well, it is a cheap horror movie. In any case, Clark and Robert now have a sense of what kind of monster they're dealing with. Later, Clark explains to David (who is horror-movie cannon fodder) how one of the shriekers has overtaken them:

0:53

CLARK: It's a three-dimensional cross-section. So it can move instantly from one place to the next, even move through walls.

DAVID: That's fuckin' great!

Clark's explanation actually makes sense. To see the idea in a simpler context, imagine a tentacled two-dimensional monster passing through a one-dimensional world. As indicated by figure 15.18, the monster's tentacles can simultaneously pass through different locations of a one-dimensional world. Presumably, Clark's four-dimensional shrieker is performing the same trick.[6]

Fig. 15.18 A two-dimensional monster can simultaneously be at different places in a one-dimensional world.

Those familiar with Edwin Abbott's classic novel *Flatland* will immediately recognize the cross-section approach to visualizing higher dimensions.[7] In Abbott's story, a three-dimensional sphere visits a two-dimensional world by passing through. As it does so, a square in the two-dimensional world observes its different cross-sections. First, a point materializes out of nowhere. This point then turns into a circle, which first expands and then shrinks back to a point.

We are aware of four movies based on Abbott's novel, of which our favorite is *Flatland: The Movie* (2007). Also charming is the claymation film *Flatlandia* (1982), made by mathematician and film maker Michele Emmer.[8]

15.6 The Hypersphere

In *Insignificance* (1985), Marilyn Monroe visits Albert Einstein to discuss the theory of relativity. Her husband, baseball player Joe DiMaggio, who likes round things, barges in on the discussion. He suspects the worst ...

56:18
MARILYN: What is the shape of the Universe?
EINSTEIN: It's not important. You've got things to discuss.
DIMAGGIO: God dammit, you tell her the shape of the friggin' Universe! I want to take her home. Tell her!
MARILYN: Please.
DIMAGGIO: Tell her!

---

[6] It would appear that the four-dimensional man in the movie *4D Man* (1959) is doing something similar, but it is never really explained. More clearly, in *The Big Bang Theory* episode "The Pants Alternative" (2010), Sheldon correctly describes how he could take his pants off over his head in a four-dimensional world. Unfortunately, he then attempts it in the regular three-dimensional world.

[7] Edwin A. Abbott, *Flatland: A Romance of Many Dimensions* (Seeley & Co., London, 1884).

[8] For a discussion of *Flatlandia*, see the article of that title in Michele Emmer, *Mathematics, Art, Technology, and Cinema* (Springer, New York, 2003), 197–201.

EINSTEIN: Well, the shape of the Universe is difficult to describe. If I were to explain it, it would come out as abstract symbols, but I'll tell you how to imagine it. Imagine a dot so small that it has zero dimensions. You extend the dot into a one-dimensional line. You turn around the line, and you've got a two-dimensional circle. And you flip the circle and you've got a three-dimensional sphere. And you got that sphere, and you turn it into the fourth dimension, and you got the shape of the Universe, but ... [Figure 15.19]

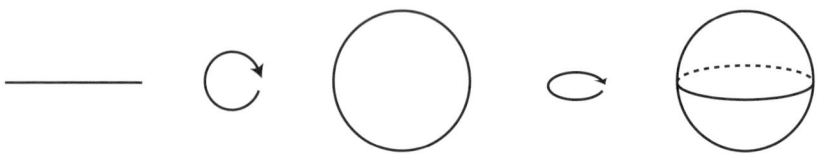

Fig. 15.19 Building higher-dimensional spheres from lower-dimensional spheres.

EINSTEIN: I know. I know now exactly how you can imagine it. Take a completely solid object, a completely solid object and twist it inside out indefinitely forever and that's the shape of the Universe.
MARILYN: Wow!
DIMAGGIO: Bullshit! I tell you what I think it is. I think it's round like everything else in nature. Like flowers, like the moon and the sun. It's all based on a circle. You know that? Like the world. I don't know what you two geniuses think the shape of the world is, but me and Columbus happen to think it's round. It's a damn lucky thing for the United States, too. For if it wasn't for Columbus we'd all be Indians. What do you think about that?

The real Albert Einstein believed that our Universe is a hollow hypersphere, that is, the higher-dimensional counterpart of a hollow sphere. The movie Einstein provides a method for visualizing a *solid* four-dimensional hypersphere (whose surface is the hollow three-dimensional hypersphere).

Although the movie people didn't get it quite right, we'll give them 7 out of 10 for getting close, plus an extra point for having it all inspired by Marilyn Monroe.[9]

Einstein's second attempt to explain things to Marilyn doesn't mean much at all. However, notice that DiMaggio's and Einstein's preferred shapes are really the same thing, just in two different dimensions.

Even more memorable than the above shape-of-the-universe scene, *Insignificance* contains an absolutely marvelous scene in which Marilyn Monroe explains the special theory of relativity to Einstein using flashlights and toy trains. Bringing us back to Einstein's four-dimensional space-time universe, that other mathematical story ...

---

[9] If we follow Einstein's instructions, but begin only with the two endpoints of the line segment, then we actually *do* obtain the hollow hypersphere.

Chapter 16
To Infinity, and Beyond!

So exclaims Buzz Lightyear in the *Toy Story* movies. However, Buzz is referring to his willingness to travel to the ends of the Universe in order to save his pal Woody. He is unlikely to be as gung ho in the realm of mathematical infinities.

It is the same with most movies. The concept of infinity is very emotive, a powerful metaphor for our physical and psychological smallness. It is understandable that almost all movies that refer to it do so in the manner of *Toy Story*, as an unexplained symbol of our tinyness, or of the vastness of our efforts and concerns. Since, then, the very purpose of referring to "infinity" is to suggest incomprehensible magnitude, it is neither a surprise nor a failing that these movies offer little true understanding.

In *Brain Dead* (1990), for example, the evil Eunice Corporation is attempting to gain the secrets lost in the brain of the mathematician Dr. Halsey. Eunice's logo is the ubiquitous symbol for infinity (figure 16.1), but the symbol here is merely a symbol.

Fig. 16.1 Infinity as a logo.

Such unexplained references to the seemingly inexplicable can be both fun and artistically effective. Mathematicians, however, have a peculiar habit of confronting the incomprehensible and making it do their bidding.[1] So it is with infinity, which has been transformed from a terrifying monster in the mathematical jungles to the most powerful tool in modern mathematics.

Some movies actually do touch upon truly mathematical notions of infinity, usually flippantly, sometimes hilariously, and just occasionally with cleverness and true understanding. Our intention here is to survey these attempts, suitably laughing and applauding as we go. In the process, in explaining how the movies get it right and wrong, we shall attempt to give some sense of just how mathematicians think of infinity, how they tamed the monster.

16.1 Mystical Musings

At the outset let's be clear that, though mathematicians may explicitly *define* infinity (and indeed they do, in various ways), this does not mean that they are completely comfortable with the concept. This is elegantly expressed by Don Pedro Valesquez in the *The Saragossa Manuscript* (1965):

2:25
DON PEDRO: I will tell you something that is based on the principles of geometry. Wanting to define an infinite number, I write a horizontal eight and divide it by one. If I want to express infinite smallness, I write a one and divide it by the horizontal eight.

All these signs give me no idea about what I want to express: infinity—greatness. In the cosmos, infinite smallness is an infinite root of the smallest part of an atom. So I am defining infinity, but without comprehending. Well, if I do not comprehend but I can define it, I am getting near poetry, which seems to be nearer to life than we suspect.

Don Pedro's reference to geometry is somewhat peculiar, but he is correct in two ways. First, to simply write the symbol for infinity explains nothing. *The Bank* (2001), for example, is a mathematically rich movie, but the use of the symbol ∞ at the end of the movie, indicating the bank's demise (figure 16.2), is simply a metaphor hiding amid the mathematics. Second, mathematicians *can* still define and use infinity, even if without full comprehension, and this process is both powerful and poetic.

16.2 Toward Infinity, but Getting Lost

Don Pedro stays safely in the realm of poetry, but things tend to get messier when a movie attempts to deal with infinity more concretely. Of course, the appearances of infinity in *Brain Dead* and *The Bank* are quite innocent, simply catchy illustrations possessing no deeper meaning. The same cannot

---

[1] For a beautiful popularization of mathematics written from this perspective, see John Stillwell, *Yearning for the Impossible* (A K Peters, Wellesley, MA, 2006).

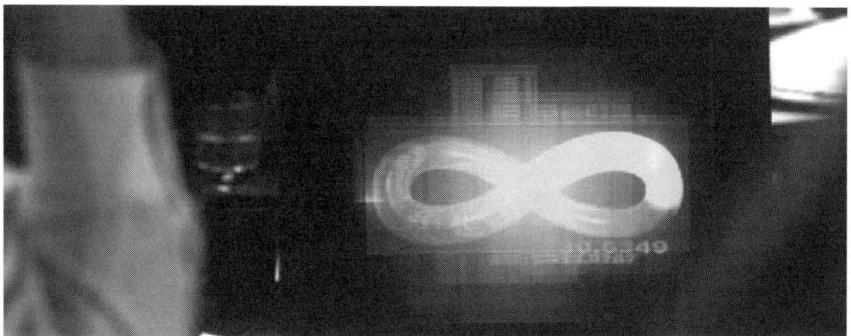

Fig. 16.2 An infinity sign signaling doom in *The Bank*.

be said for the following exchange in *Alien Hunter* (2003), in which, sad to say, Julian (James Spader) is our hero:

1:16
KATE: You know how unlikely that is.
JULIAN: How unlikely? What are the odds?
SHELLEY: 99.999 to the infinite.
JULIAN: But not 100.

16.3 Toward Infinity, and Almost Getting There

The above scene from *Alien Hunter* contains some pseudomathematics, but there is no genuine attempt to deal with infinity. To do so, clearly one must carefully consider exactly what "infinity" means.

The simplest approach to mathematical infinity, the notion of an *infinite process*, arises in *Forbidden Planet* (1956). At one point, Dr. Morbius is explaining to Lieutenant Ostrow and Commander Adams the amount of energy the Krell aliens were able to produce (figure 16.3):

0:58
MORBIUS: May I draw your attention to these gauges all around here, gentlemen. Their calibrations seem to indicate that they are set in decimal series, each division recording exactly 10 times as many Amperes as the one preceding it. Ten times ten times ten times ten, on and on and on, row after row, gauge after gauge. But, there is no direct wiring that I can discover. However, when I activate this machine it registers infinitesimally, you see down there in the lower left-hand corner. And, then, when I activate the educator here, it registers a little bit more.
OSTROW: But this much is negligible, the total potential here must be nothing less than astronomical!

MORBIUS: Nothing less. The number ten raised almost literally to the power of infinity.

Fig. 16.3 Dr. Morbius in *Forbidden Planet*.

Here we have the idea of unbounded growth, multiplying by ten with each gauge, creating the sequence

$$10^1, \ 10^2, \ 10^3, \ldots$$

Of course, the Krell have a definitively *finite* planet, the sequence of meters must eventually end, and Dr. Morbius's suggestion of "almost literally" reaching infinity can make no real sense. Nonetheless, the meters give a sense of *potential infinity*, the idea of unbounded growth, that we can *imagine* the meters going on forever.[2] This parallels the growth of the natural numbers and is *exactly* what we mean by saying there are infinitely many of them.[3]

16.4 Toward the Infinitely Small: Romantic Zeno

This idea of infinity as a process can be considered in exactly the same manner for the infinitely small. For a romantic view of this, we turn to *I.Q.* (1994), where Catherine (Meg Ryan) is explaining one of Zeno's paradoxes to Ed (Tim Robbins) (figure 16.4):

---

[2] Mathematically, we would be describing just as valid a potential infinity if the meters increased according to the sequence 1, 2, 3, ... Of course, this slower growth is far less suggestive of the psychological sense of infinity.

[3] A similar discussion, concerning the infinite collection of prime numbers, appears in *The Mirror Has Two Faces* (1996); see chapter 11.

0:49

CATHERINE: You can't get from there to here because you always have to cover half the remaining distance (figure 16.4). Like from me to you, I have to cover half of it. But see I still have half of that remaining—so I cover half that—And there is still half of that left, so I cover half of that—and half of that, and half of that, and half of that. And since there are infinite halves left, I can't ever get there.

Ed then remarks that Catherine has nonetheless reached him. And, undeniably together, they dance.

Fig. 16.4 And half of that . . .

16.5 To Infinity: Are We There Yet?

But what about *actually* getting there? The idea in the previous scenes is not that any one number is of infinite extent, but of the process, the potential to grow larger (or smaller) without bound. But to actually get there?

This is the notion of *completed infinity*, of somehow capturing the totality of the process. In Catherine's case, this would amount to not simply being as close as desired to the desirable Ed, but to being in the exact same spot. In the case of Dr. Morbius, we would have to somehow deal with ∞ as a number, as in itself a possible magnitude.

Mathematicians do indeed treat ∞ as a number, including performing (careful!) arithmetic with it, but this notion is not well known or understood outside of the mathematical world.[4] Nonetheless, the idea of ∞ as a number is quickly and humorously but accurately referred to in *Bill & Ted's Bogus*

---

[4] As simple examples, it is clear that we should have $\infty + \infty = \infty$, and $\infty + 0 = \infty$. Difficulties arise if we try to define quantities such as $\infty - \infty$, so we simply don't. This pick-and-choose approach is not uncommon. For example, though 0 is a bona fide number, we usually make no attempt to define $\frac{0}{0}$.

Journey (1991). Bill and Ted wind up in Hell and are confronted by Colonel Oats, ordering them to do push-ups:

0:39
OATS: Now get down and give me—infinity!
BILL: Dude, there's no way I can do infinity push-ups.
TED: Maybe if he lets us do 'em girlie style.

16.6 Fishing with a Really Big Net

An alternative aspect of the infinity of the natural numbers is to simply gather them all up and to regard them together as a collection, what mathematicians refer to as a *set*. This is perhaps a more familiar notion than that of ∞ as a number, but the mathematics of infinite sets quickly takes on a paradoxical air. This is introduced very nicely in *Infinity* (1996), the biopic about the physicist Richard Feynman. In one scene, Feynman is using infinity to capture young Henry's imagination:

1:47
FEYNMAN: Hey Henry. Did you know that there are twice as many numbers as numbers?
HENRY: Oh, come on, Dick!
FEYNMAN: There are. Let me show you. Name a number.
HENRY: One million.
FEYNMAN: Two million.
HENRY: Twenty-seven.
FEYNMAN: Fifty-four. Now you try it. Six.
HENRY: Twelve!
FEYNMAN: Six million.
HENRY: Twelve million!
FEYNMAN: Light beginning to dawn?

Feynman is suggesting a particular *correspondence* between natural numbers, each number being associated with its double.

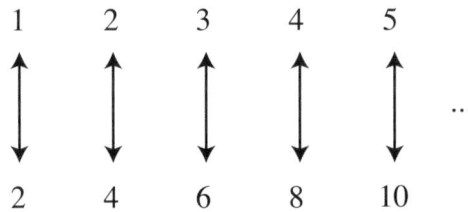

Fig. 16.5 There are as many even numbers as numbers.

What can we see from this? Most simply, as Feynman goes on to remark, the correspondence demonstrates that there is no largest number, since the double will always be larger.[5] But Feynman's original suggestion is that we can see something much more intriguing, that there are twice as many numbers as numbers. How can we make sense of this?

Feynman's correspondence makes clear that *there are exactly as many even numbers as natural numbers:* the pairing shows that the second row in figure 16.5 has exactly as many things in it as the first row. This is already peculiar,[6] but we can go further.

We can just as easily see that there are exactly as many *odd* numbers as numbers (pairing $1 \leftrightarrow 1$, $2 \leftrightarrow 3$, $3 \leftrightarrow 5$, and so on). Then, since the collection of natural numbers is exactly the collections of even and odd numbers combined, there are twice as many numbers as numbers, just as Feynman promised.

Does this make any sense? Perhaps surprisingly, it does; one simply has to accept that the mathematics of infinite collections behaves differently, with an inevitably paradoxical feeling to it. We already stumbled across this in *The Mirror Has Two Faces* (1996), when the mathematician Greg Larkin confronts his student:

1:45

GREG: Don't you know that it is possible to remove an infinite number of elements from an infinite set and still have an infinite number of elements left over?

Greg is thinking of prime numbers, but in terms of our setting, an example would be to remove the infinitely many even numbers, leaving behind the infinitely many odd numbers.[7] Unfortunately, in *Infinity*, Feynman cops out by making his much simpler and less controversial observation, never returning to his initial teaser.

The fascinating mathematics of infinite correspondences was cleverly described a century ago by mathematician David Hilbert. Hilbert imagined a hotel with infinitely many rooms and infinitely many guests. He then considered various mathematical correspondences by imagining new guests arriving at the hotel and having them accommodated by shuffling around the original

---

[5] The feature-length cartoon *The Phantom Tollbooth* (1970) considers infinity in a similar manner: when Milo visits Digitopolis, its ruling mathemagician convinces Milo of the infinity of numbers by adding 1 again and again. Earlier, the mathemagician sings a version of Zeno's paradox, suggesting that Milo keep dividing his troubles in 2 until they disappear!

[6] That is, we have demonstrated that an infinite set can be the same size as one of its subsets. This is in fact a much more common introduction to the weirdness of infinite sets, with Feynman's "twice as many numbers as numbers" observation usually an afterthought.

[7] Consideration of the size of these sets suggests the numerical equation $\infty - \infty = \infty$, to be contrasted with the "natural" $\infty - \infty = 0$. This demonstrates the fundamental danger with the arithmetic of ∞, referred to earlier. Subtraction of ∞ from ∞ can mean many different things. So, since we don't have just one meaning for $\infty - \infty$, we have no better choice than to leave it undefined.

guests. Hilbert's playful approach to infinity was brilliantly brought to life in the stunning education film *Hotel Hilbert* (1996).

16.7 Infinity Pays Its Way

Much more than a beautiful concept, infinity is an actively used tool in modern mathematics. Again, there are movies that give some indication of this, but we shall begin with the silly. In *Las Vegas Weekend* (1986), the "math wizard" Percy Doolittle is describing his theories to his new gambling partners:

0:21
PERCY: You see, my work involves a new system of mathematics, the mathematics of infinity, a place where time and space stand still. And they don't stand still so much as they merge into one piece of time—space.—So I calculate what happens out there in that random space, infinity, and it's funny but it's very much like predicting gambling and what cards are coming up next. Because it is after all the same phenomenon, random events.

This is "completely gobbledegooky," as Percy's PhD supervisor aptly summarizes, although it sits well enough in a dumb comedy about gambling. A more serious movie with a more serious, if not completely successful, approach to mathematics is *Moebius* (1996). The story involves a subway train which has somehow disappeared without trace from the network. Daniel Pratt is a mathematician brought in to consider the event:

0:49
DANIEL: I'm not sure but I think it has integrated the whole system at such a high level that I don't even know how to calculate it. I suppose it's become infinite. If I'm right, gentlemen, we can conclude that the system is working like a Möbius band.

Moebius contains many such remarks upon the newly infinite nature of the subway network. However, even accepting the fantasy world of this movie—and the idea that an introduced Möbius band might muck up the connectivity of the subway network is impossible but delightful—the remarks about infinity make no sense.

The Möbius band is fascinating due to its one-sidedness, and its peculiar characteristics are often emphasized by drawing the Möbius band in the shape of ∞, as did the Eunice corporation in *Brain Dead* (figure 16.1), and as is indicated in the titles of *Moebius* (figure 16.6).[8] But there is nothing especially

---

[8] One source of encouragement for this association of the Möbius band with infinity is the manner in which paper Möbius bands will naturally form the shape of the symbol ∞. This is charming but is of no particular mathematical importance.

"infinite" about it. True, one can travel forever on a Möbius band, making loop after loop, but exactly the same can be said for a circle or a cylinder.[9]

Fig. 16.6 An ∞-shaped Möbius strip in the titles of *Moebius*.

Even those movies that explicitly deal with mathematics usually spend little time on the specific notions and applications of infinity. *A Beautiful Mind* (2001), for example, treats some higher mathematics with clarity and intelligence, and infinity is implicit in much of this. However, the only explicit appearance of infinity is an unexplained infinite sum in John Nash's window doodlings, and an infinity formed by the path of Nash's bicycle (figure 16.7).

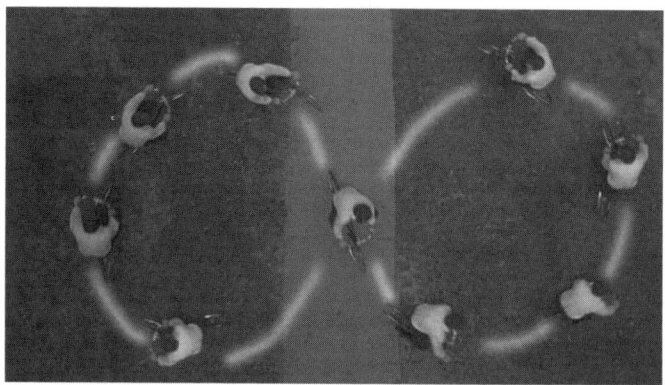

Fig. 16.7 John Nash rides to ∞ in *A Beautiful Mind*.

The Bank (2001) offers a little more (figure 16.8). When the mathematician Jim Doyle employs dynamical systems in an attempt to master the stock market, he indicates what he needs to succeed:

0:10
JIM: This mathematics allows us to predict almost anything. If we can predict this [he draws an ∞], *then predicting the stock market is easy.*

Similar is Professor Benson in *Battle of the Worlds* (1961), explaining how he noticed the outsider, a moonlike object heading straight for Earth:

[9] For explicitly stated confusion on this point, see the writer-director's commentary on *Pleasantville* (1998).

Fig. 16.8 Encountering infinity in *The Bank*.

0:10

PROFESSOR BENSON: You and the others have to see and hear before you can know. I have one advantage over all of you: calculus.—Oh, so you didn't see it until just before dawn. And didn't anyone of you notice the change of position of the two outer planets?—Infinitesimal! It merely heralded the arrival of the outsider.

16.8 Pretty Patterns, Pretty Pi

Professor Benson declares the power of the concept of infinity through the power of the calculus but, as for all the above scenes, he suggests such power while providing no detail whatsoever. This is hardly surprising: the beauty and appeal of higher mathematics are anything but easy to convey, and the movie-going public is hardly primed to receive such material. So even a mathematically ambitious movie-maker is advised to employ more familiar, more accessible and more visually appealing material.

π (1998), as we saw in chapter 4, is one movie that does so. This movie treats mathematical concepts, and infinity in particular, with thought and integrity. This is made clear in the very first scene, where mathematician Max Cohen calculates $\frac{73}{22}$, rhythmically reciting the decimal expansion as he disappears down the stairwell. The vision of Max descending as his voice trails off gives a vivid picture of the infinite decimal expansion of $\frac{73}{22}$.

Of course, though this representation of $\frac{73}{22}$ is infinite, the fraction is not itself an infinite creature. The same cannot be said for π, the movie's cen-

Fig. 16.9 $\frac{73}{22} = 3.31818181818\ldots$

tral, irrational, character: the opening titles show π superimposed upon the beginning of its decimal expansion.[10]

This raises an interesting issue about dots and the representation of infinite processes. If we write $\frac{73}{22} = 3.31818\ldots$, we know exactly what the dots mean: just keep repeating the "18" forever. However, if we write $\pi = 3.141592\ldots$, it is not at all clear what the dots mean, what is meant to come next other than "what works." It is this very mystery, the seeming patternlessness in the infinite decimal expansion of π, which is at the heart of the movie, Max's search for patterns in this apparent randomness.

Of course, there are other, more natural ways to view π. At one point we hear Max's thoughts, as he considers his old professor Sol. He contrasts the geometric simplicity of π with the complexity and mystery of its decimal expansion.

0:12

MAX: We see the simplicity of the circle, we see the maddening complexity of the endless string of numbers, three point one four, off to infinity.

As Max says this, he writes the decimal expansion of π, together with the familiar equations for the area and circumference of a circle (figure 16.10). It is these very natural geometric ideas that explain why we even bother with π; they contain a simplicity and beauty much less likely to drive Max crazy. Indeed, the infinite nature of π has seemingly disappeared altogether.

However, the discerning eye can still spot infinity, peeking from underneath this geometric cloak of simplicity; the question is, how does one *define* the area (similarly, the circumference) of a circle? Certainly nothing as simple as "length times width," as we declare for a rectangle, will work. But, in the end, area can really be nothing more than "length times width" used in more and more complicated settings. Pondering this, one is inevitably led to the idea of the circle itself as a completed infinity, as an infinite-sided polygon (figure 16.11).

---

[10] Sadly almost all of the expansion is incorrect; see chapter 4 for details. See also *Red Planet Mars* (1952), where the fundamentally infinite nature of the decimal expansion of π is used to communicate with the "Martians."

Fig. 16.10 $\pi = 3.14159265\ldots$

Fig. 16.11 A circle as a limit of a sequence of regular polygons.

And so, infinity reappears in π, albeit in a much more beautiful and natural form than that of the decimal expansion.[11]

16.9 Golden Infinity

Infinity makes a last, lovely appearance in π, when Max draws and describes a *golden rectangle*. As Max indicates, a golden rectangle is a rectangle from which a square can be removed, to leave a rectangle of the same proportions. There is nothing glaringly "infinite" about this construction, but Max notes that this process of "squaring" can be repeated, over and over, creating an infinite nested sequence of golden rectangles (figure 16.12).

---

[11] Even if we consider numerical expressions, there are much more elegant ways to represent π than by its decimal expansion. For example, the following is π as an infinite product (discovered by seventeenth century mathematician John Wallis):

$$\pi = 2 \cdot \frac{2}{1} \cdot \frac{2}{3} \cdot \frac{4}{3} \cdot \frac{4}{5} \cdot \frac{6}{5} \cdot \frac{6}{7} \cdots .$$

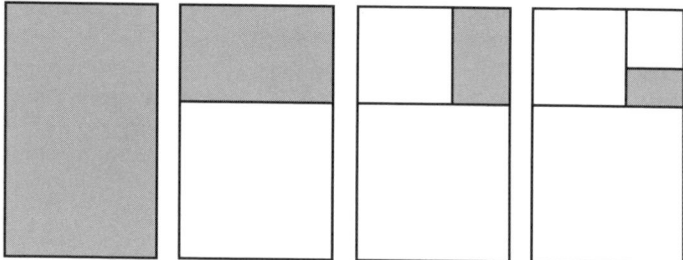

Fig. 16.12 Constructing smaller and smaller golden rectangles.

On the other hand, this possibility is hardly unique to the golden rectangle. In *Donald in Mathmagic Land* (1959), for instance, there appears a beautiful infinite nest of regular pentagons and pentagrams (figure 16.13).

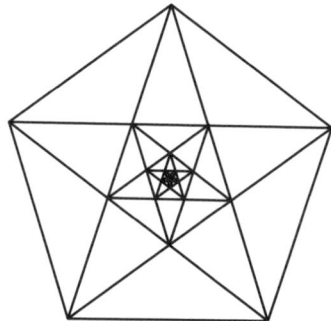

Fig. 16.13 Nested pentagons and pentagrams.

In both cases, one can stop there, simply admiring the beautiful diagrams. But we can go further, applying truly infinite arguments to deduce that the golden rectangle is a truly infinite creature.

16.10 A Golden Argument

The *golden ratio* ϕ is defined to be the length divided by the width of a golden rectangle. It is indeed both golden and a ratio, but there is more: *the golden ratio is irrational*. That is, although ϕ is defined as a ratio, we cannot write $\phi = \frac{A}{B}$ as a fraction, as the ratio of two natural numbers A and B.

How can we see this? Suppose that ϕ *were* a fraction. That would mean there was a golden rectangle with natural number sidelengths A and B (figure 16.14).

But then, the next golden rectangle would have natural number sides (of lengths B and $A - B$). And, so would the next, and the next, and the next.

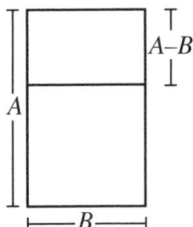

Fig. 16.14 A, B, and $A - B$ are all natural numbers, with $A > B > A - B$.

So, we would have an infinite, descending sequence of natural numbers, and that's impossible!

16.11 Poetic Summation

Finally, we return to the *The Saragossa Manuscript*, where Alfonse Van Worden criticizes Don Pedro's use of a mathematical analogy. Don Pedro responds, pithily describing the power of the mathematics of infinity, and of mathematics in general:

2:29
DON PEDRO: Something like quotients which can be divided infinitely.
VAN WORDEN: I am a captain of the guard. Not a philosopher. Your mathematics is just dead numbers.
DON PEDRO: Señor, this zero, plus and minus one, gave Newton and Archimedes power equal to the gods.

Chapter 17
Problem Corner

Thérèse is a mathematical prodigy in *Antonia's Line* (1985). We see her as a young girl, her precociousness on display:

ANTONIA: Since when can you do additions?
THÉRÈSE: Since I was three.
ANTONIA: What is 147 times 48?
THÉRÈSE: 7056, of course.—Square root is 84. The square is 49,787,136.
ANTONIA [to Thérèse's parents]: I regret to have to tell you that your daughter is not normal.
THÉRÈSE: I am a wunderkind.

In this chapter we collect some of the problems, from math competitions and the like, that have confronted movie characters. See also chapter 7 ("Escape from the Shrinking Square") for a movie full of puzzles. For the benefit of anyone who may not quite be a Thérèse-like wunderkind, we also supply some answers and some hints.

17.1 Problems for Wizkids, and a Wizdog

We begin with some problems from *Little Man Tate* (1991), a movie about wizkids. The mathematical stars are little Fred and mathemagician Damon. The first few problems are from a math competition in which both participate.

Problem 1: How many minutes are there in 48 years?

Answer (courtesy of Damon): 25,228,800 minutes—151,368,000 seconds.

It seems Damon is assuming that there are 365 days in a year, which is a trifle odd. Even then, the number of seconds should be 1,513,728,000, but nobody seems to notice.

Problem 2: How many factors are there in 3067?

Answer (Damon again): Come on guys. There are no factors of 3067. The number is prime. Somebody, for God's sake, challenge me!

Problem 3: How about giving me a number, that when divided by the product of its digits the quotient is three, and if you were to add 18 to this number the digits would be inverted?

Answer (Fred): 24.

Assuming the number we're looking for has two digits, we can write it as $10X + Y$, with X and Y natural numbers from 0 to 9. Then the information provided gives us two equations for X and Y, and it's easy to show that the only integer solution is $X = 2, Y = 4$. One can also look for solutions with more than two digits: it's trickier, but it can be shown there are no further solutions to be found.

Problem 4: What is the cube root of 3,796,466?

Answer (Fred): 156.

Again, there is a mistake. For Fred's answer to be correct, the quizmaster should have asked for the cube root of 3,796,416.

The following is our favorite problem in this movie.

Problem 5 (Teacher) How many of the numbers 1, 2, 3, 4, 5, 6, 7, 8, 9, and 10 are divisible by 2? [Figure 17.1]

Fig. 17.1 How many of these numbers are divisible by 2?

Answer (Fred): Um, all of them.

And one more problem from the math competition.

Problem 6: What number has the following peculiarity. If its cube were added to five times its square and from the result 42 times the number and 40 is subtracted, the remainder is nothing.

Answer (Fred): 5.

If we let X stand for the number we're after, then the information provided amounts to the equation $X^3 + 5X^2 - (42X + 40) = 0$. One solution to this equation is Fred's answer, $X = 5$. However, since it's a cubic equation, we expect two more solutions. Given that we know one answer, it is not hard to factor $(X-5)$ into the cubic, and then use the quadratic formula to obtain the two answers that Fred missed: $X = -5 + \sqrt{17}$ and $X = -5 - \sqrt{17}$. Possibly Fred only concerned himself with finding an integer solution. However, after making a fool of his teacher with the previous problem, it is only fair that we nitpick Fred here.

Next, we have the questions from the School Mathletes State Championship in *Mean Girls* (2004).

Problem 7: Twice the larger of two numbers is three more than five times the smaller, and the sum of four times the larger and three times the smaller is 71. What are ... ?

Answer: 14 and 5.

Writing X and Y for the two numbers, it is easy to set up and solve the simultaneous equations.

Problem 8: Find an odd three-digit number whose digits add up to 12. The digits are all different, and the difference between the first two digits equals the difference between ...

Answer: 741.

The quizmaster doesn't mention it, but 147, 345 and 543 also work.

Problem 9: Find the limit of this "equation":

$$\lim_{x \to 0} \frac{\ln(1 - x) - \sin x}{1 - \cos^2 x}.$$

Answer: The limit doesn't exist!

This is the problem Cady (Lindsay Lohan) must solve to grab her school the title. She gets the answer correct. However, except for the advice to herself to stop thinking about boys, she gives no clue how she does it.

Here are two nice puzzles from *Eustice Solves a Problem* (2004).

Problem 10: How many ways are there of walking up a flight of 10 stairs if you take either 1 or 3 steps with each step?

Answer: 28.

This problem is closely related to the Fibonacci numbers. Writing S_N for the number of ways of walking up N stairs by taking 1 or 3 steps at a time, what we're asked for is S_{10}. Taking it slowly, it is obvious that there is only one way to climb one or two steps, so $S_1 = 1$ and $S_2 = 1$. Given three steps, we can either take them one at a time or do a 3-step leap, and so $S_3 = 2$.

For N beyond 3, any method of getting up N steps will first arrive at the $(N - 1)$st step or the $(N - 3)$rd step. This leads to the key equation $S_N = S_{N-1} + S_{N-3}$. We can then churn out the sequence up to the 10th term: 1, 1, 2, 3, 4, 6, 9, 13, 19, 28. Note that, if instead we were allowed to take 1-step and 2-step moves, the key equation would be $S_N = S_{N-1} + S_{N-2}$, which would give the Fibonacci numbers.

Problem 11: In a bakery the ratio of pies to cakes is 2:3 and the ratio of cakes to scones is 8:1. What is the ratio of pies and cakes to scones?

Answer: 40:3.

Letting P, C, and S stand respectively for the number of pies, cakes, and scones, the information provided is $\frac{P}{C} = \frac{2}{3}$ and $\frac{C}{S} = \frac{8}{1}$. Then $\frac{P}{S} = \frac{16}{3}$, and so $\frac{P+C}{S} = \frac{8}{1} + \frac{16}{3} = \frac{40}{3}$.

The movie *Tom & Viv* (1994) is about T. S. Eliot and his supposedly crazy wife Vivienne. It features a couple of clever math puzzles. In fact, some (truly crazy) people decide that if Vivienne cannot solve the second and third puzzles below, then this means that she is crazy and needs to be committed to an insane asylum.

Problem 12: A lady when asked her age replied that she was 35, not counting Saturdays and Sundays. What was her real age?

Answer: 49.

The lady is counting only $\frac{5}{7}$ of the days, and so her pretend age should be multiplied by $\frac{7}{5}$.

Problem 13: Rupert takes his friends to the opera. Rupert is sitting next to Charles and on his left, Daphne immediately on Charles's right, and Clarissa sits somewhere to the left of Daphne. Can you put them into their correct order?

Answer: From left to right, the order is Clarissa, Rupert, Charles, Daphne.

Problem 14: A greasy pole is 10 yards high. A little brown monkey wishes to climb the pole. The monkey climbs 3 yards a day, and at night he slips back 2 yards. How many days will it take him to reach the top?

Answer: 8.

After the seventh night, the monkey will be 7 yards up. So he'll make it to the top on the eighth day, and then slip back down 2 yards on the eighth night.

In the Kung Fu movie *Brave Archer III* (1981), the mysterious Auntie Ying is teased with a problem.

Problem 15: A number that divided by 3 leaves 2, divided by 5 leaves 3, and divided by 7 leaves 2. What is the number?

Answer: 23, though Auntie Ying doesn't answer.

Once we've checked that 23 works, notice that adding $3 \times 5 \times 7 = 85$ won't change the remainders. So 108 also works, as does $23 + 85N$ for any natural number N.

Bingo (1991) is the story of a very smart dog. In one scene, he helps his owner Chuckie with his homework.

Problem 16: What is the square root of 9?

Answer: Woof! Woof! Woof!

17.2 Math Quiz for Mortals

Here are some problems from the math competition in *Lambada* (1991). Some supposed no-hopers come through on the big day.

Problem 17: You plotted the position of a star in relation to the background stars six months ago and again last night. The angle of shift in the star's position is eight-tenths of a second. How would you determine the distance of this star from the Earth?

Answer: "With a big tape measure. [*Laughter*] Okay, okay. But if I couldn't find no tape measure, I guess I'd have to use the trigonometric parallax. See, I'd start with a right triangle formed by the Sun, the stars and the position of the Earth last night. That makes my angle of interest 0.4 seconds, and the far side is the distance from the Earth to the Sun in one astronomical unit. The near side, with the length we don't know, is the distance from our solar system to the star. Now you measure the ratio far side over near side for a right triangle when the angle of interest is 0.4 seconds. Uh, somethin' like that."

The suggested solution is not precisely correct, although it is probably close enough in practice. The star, the Earth, and the Sun will not in general form a right triangle, and the angle of interest will not in general be 0.4 degrees. And, because of the Earth's elliptical orbit, the Sun will be at a focal point of the orbit rather than at the center. However, because the star is so far away in comparison to the Sun-Earth distance, treating the triangle as a right triangle is a quick and reasonable method for obtaining an estimate of the distance to the star.

Problem 18: Describe the Cartesian coordinate system.

Answer: It's a system where points on a plane are identified by pairs of numbers that represent distances from two perpendicular lines.

Problem 19: The abscissa of a point in the plane is the distance of the point from the x-axis. The ordinate is the distance of the point from the y-axis. Given an equation $2x + 3y = 12$, what are the coordinates of the solution?

Answer: "You got the question wrong." "Young lady, if you can't answer the question, leave the microphone." "No, see you reversed them. The abscissa is the vertical line from the point in the plane from the y-axis."

No one is doing very well here, with both getting the definitions wrong: if the coordinates of a point are (x, y) then the abscissa is x, i.e., the distance from the point to the y-axis, and the ordinate is y. Anyway, the no-hoper answers with one of the infinitely many correct solutions.

In *Little Big League* (1994) the Minnesota Twins baseball team has to help their teenage manager finish his homework before the big game. Absolutely hilarious.

Problem 20: If Joe can paint a house in three hours and Sam can paint the same house in five hours, how long does it take for them to do it together? [Figure 17.2]

Answers: "You never said this was a word problem." "What color paint?" "It's simple, 5 times 3, that's 15" "No, no, no. Look, it takes 8 hours, 5 plus 3." "Check it out, there is one, two of them. It only takes 4 hours." "I should know this, my uncle's a painter." "Why don't they just get a house that's already painted." "You know, maybe there is no answer, maybe this is one of those trick questions. Did you ever think of that?" "I don't know, I mean 8 sounds good to me." "Fellows, fellows, fellows, fellows, fellows. The chalk, if you please. Thank-you. Using the simple formula A times B over A plus B, we arrive at our answer of $\frac{3 \times 5}{3+5}$, equals one and seven eights." "Wow, are you sure." "Oh ho, but of course, my diminutive leader. Long have I been familiar with the exactitudes of the mathematical world."

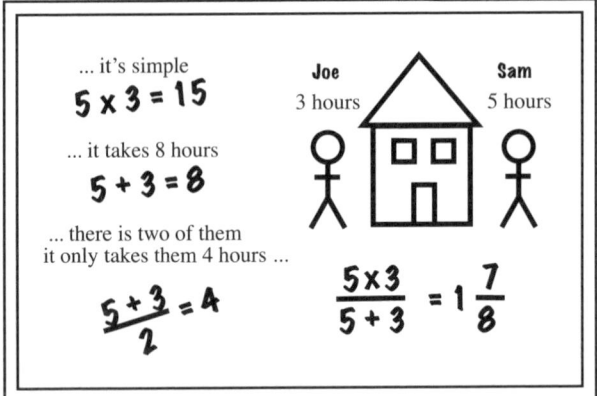

Fig. 17.2 What color paint?

Here is a problem attempted by one of the protagonists in *Wish Upon a Star* (1996).

Problem 21: A cleaning solution is made up of three chemicals. A, B, and C. There are equal amounts of A and B and four times as much C as there is A. What percentage of the bottle is full of C?

Answer: $B = A$ and $C = 4A$. So $A + B + C = A + A + 4A = 6A$. Therefore, your percentage is $\frac{4A}{6A}$, the As cancel, $\frac{4}{6}$ is converted to $\frac{2}{3}$. Therefore, the percentage is 66.6%.

17.3 Devilish Problems

The first devilish problem appears as homework in *Bedazzled* (2000), with Elizabeth Hurley playing the devil.

Problem 22: $x^n + y^n + z^n$. Solve for $n > 2$. Show your work!

Answer: See chapter 14 ("Pythagoras and Fermat Go to the Movies").

In *Diabolique* (1955), the teacher tests her students with the following geometry problem.

Problem 23: The area of a hexagon in relation to the radius of the circumference?

Answer: In the movie, the answer is given as $6AB \times \frac{1}{2}OH$, with AB equal to the radius and OH the altitude of one of the equilateral triangles making up the hexagon (figure 17.3). This is correct as far as it goes, but does not really answer the question. The precise answer is $\frac{3\sqrt{3}}{2}AB$.

In 1996, *Diabolique* was remade, again titled *Diabolique*. Sharon Stone appears as the math teacher, and there is a corresponding classroom scene. As a sign of the times, the problem posed is of a slightly different caliber:

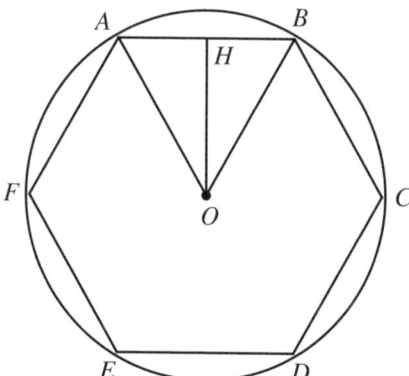

Fig. 17.3 Calculating the area of a regular hexagon in *Diabolique*.

Problem 24: 10 plus x when x equals 3 is?

Any reader who struggles with this problem should feel free to contact us for the solution.

17.4 Crazy Problems for Extra Credit

Here is a very tough problem, although it doesn't phase the robot Brainman in the *Thunderbirds* episode "Sun Probe" (1965).

Problem 25: I want you to calculate the following equation. What is the square root to the power of 29 of the trigonometric amplitude of 87 divided by the quantitative hydraxis of 956 to the power of 77?

Answer: 45,969, apparently.

In the teen slasher flick *Cutting Class* (1989), Paula and her math teacher are set a word problem to solve. The answer will inform them whether they can safely exit by door number 1 or door number 2.

Problem 26: Which door? Get it right or die! A train leaves Chicago at 8, heading east. Another train leaves Boston heading west at 8. At what time do the trains collide? X equals 1 or 2?

Answer: Not $X = 1$. The killer gruesomely informs the math teacher that he got it wrong, because he forgot to allow for the difference in time zones.

Finally, here is a problem from the Abbott and Costello movie *Buck Privates* (1941).

Problem 27: You're 40, she's 10. You're four times as old as that girl. Now you couldn't marry her, so you wait 5 years. Now the little girl is 15, you're 45. You're only three times as old as that little girl, so you wait 15 years more. Now the little girl is 30, you're 60. You're only twice as old as that little girl. How long do you have to wait until you and that little girl are the same age?

Answer: "What kind of question is that?" "Answer the question." "That's ridiculous." "What's ridiculous?" "If I keep waiting for her she'll pass me up." "What are you talking about?" "She'll wind up older than me."

Chapter 18
Money-Back Bloopers

It's pretty hard for a movie with math to totally avoid bloopers. However, some are simply unforgivable, or are forgivable and just really, really funny. In this chapter, we've collected some of the best.

18.1 Boosting the Computer

In the *Star Trek* episode "Court Martial" (1967), Captain Kirk comes up with a brilliant idea:

CAPTAIN KIRK: Gentleman, this computer has an auditory sensor. It can, in effect, hear sounds. By installing a booster we can increase that capability on the order of one to the fourth power.

18.2 Playing the Percentages

In *Alien Hunter* (1987), the mathematician Julian is calculating the chances:

SHELLY: It still could be hidden in the protein.
KATE: You know how unlikely that is.
JULIAN: How unlikely? What are the odds?
SHELLY: 99.999 to the infinite.
JULIAN: But not 100.

In the second episode of the TV series *FlashForward* (2009), the FBI is pondering the chances of everyone on the planet blacking out at exactly 11 a.m. (Pacific time):

AGENT: There are 60 minutes in an hour, 60 seconds in a minute. So the chances of something happening at exactly the top of any hour are 1 in 3600.

It's reassuring to know that the FBI are on the case.

In *The Bridge on the River Kwai* (1957), Major Warden is deciding whether Commander Shears should take some practice parachute jumps before the big mission:

WARDEN: They say if you make one jump you've only got 50% chance of injury, two jumps 80%, and three jumps you're bound to catch your packet. The consensus of opinion is that the most sensible thing for Major Shears to do is to go ahead and jump and hope for the best.
SHEARS: With or without a parachute?

They're in the middle of a war, so one can understand they'll go with rough and ready calculations. However, assuming that the chance of getting injured on any jump is independent of the other jumps, the percentages should be 50%, 75%, and 87.5%. And, one would hope that the benefit of practice is that the later percentages would be lower.

In *The Arrival* (1996), Ilana and her fellow climate scientist are discussing the increases in certain gases:

ILANA: 700% increase over the last five years. How can that be?
SCIENTIST: My very question. Just how accurate is this data you sent us, Ilana?
ILANA: Well, that's hard to say. We cobbled it together from ground stations, weather balloons, Uncle Earl's aching corns. Some of your own numbers are in there.
SCIENTIST: Well, you've obviously got some ratty data.
ILANA: We checked this as best we could.

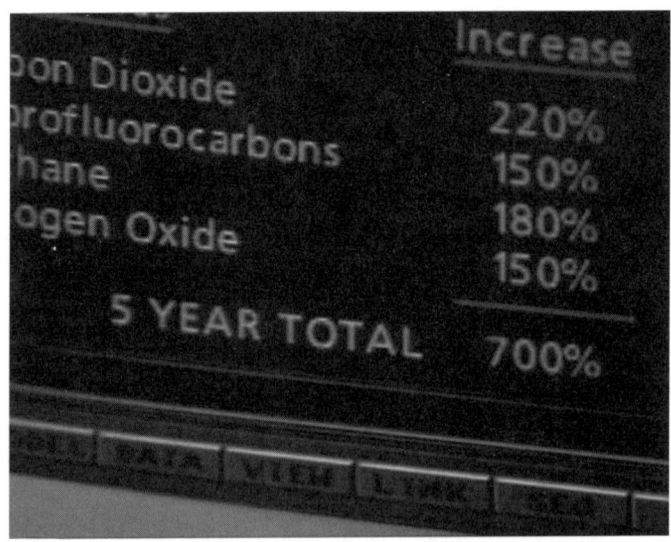

Fig. 18.1 Some brilliant combining of percentages in the *The Arrival*.

Their data may well be ratty, but not nearly as ratty as their method of combining percentages (figure 18.1).

18.3 The Curse of Pi

Apparently, almost any movie that tries to nail down more than a few digits of the decimal expansion of π is doomed to fail. In *Donald in Mathmagic Land* (1959) the birdlike pi creature recites: "π is equal to 3.14159265389747, etcetera, etcetera, etcetera." Oddly, the last two digits are wrong, and it should be 3.141592653589**793**... In *The Virgin Suicides* (1999), a classroom displays what are supposed to be the first forty-seven digits of π; sadly, things already go wrong after the eleventh digit. The title sequence of the movie π (1998) has thousands of digits scrolling across the screen, clearly meant to be digits of π, but only the first nine digits are correct. And the grand prize goes to *Never Been Kissed* (1999), with an earth-shattering three correct digits (figure 18.2):

Fig. 18.2 π (more or less) in *Never Been Kissed*.

Another interesting attempt occurs in Kate Bush's song "Pi," from her album *Aerial* (2005). It's a beautiful song, but her attempt to recite the first 138 digits of π goes awry:

3.
1415926535 8979323846 2643383279 5028841971 6939937510
5820974944 5923078164 0628620<u>899</u> <u>8628034825</u> <u>3421170679</u>
82148<u>0</u>8651 3282306647 0938446095 5<u>0</u>58223.

All the underlined digits are missing. But it's a weird song, so maybe it's just that the deeper meaning was lost on us.

It seems that π can also curse movies in other ways. In *St. Trinians* (2007), there begins a very funny scene where Peaches answers that the volume of a sphere is "quite loud." Unfortunately, the intended humor is overshadowed by the unintended humor, when everybody agrees that the correct answer is actually πr^3. This exact same blooper, along with the traditional blooper $\pi = \frac{22}{7}$, appears in the Pyecraft episode of *The Infinite Worlds of H. G. Wells* (2001); see figure 18.3.

Fig. 18.3 Some good and bad π.

18.4 Prime Problems

In *Contact* (1997), the heroine Ellie (Jodie Foster) explains what a prime number is:

ELLIE: Prime numbers. That would be integers that are only divisible by themselves and 1.

Almost. Except that Ellie should have specified integers greater than 1. After all, it is an important convention not to count the number 1 among the prime numbers. True, this is nitpicky, but sometimes it is important to get this right.

For example, in *The Core* (2003), the hero is told that a secret message can be decrypted using prime numbers. He immediately types in 1, 2, 3, 5, 7, 11, 13, 17. Voilà, it works! Luckily both he and the person who encrypted the message are under the impression that 1 is a prime number.

Even experts can get this wrong. Here's math professor Gregory Larkin in *The Mirror Has Two Faces* (1996):

GREG: My number here is 01712577355, all prime numbers, by the way.

Well, there are definitely a few that are not . . .

In *The Music of Chance* (1993), the weird duo of Flower and Stone explain how they chose their winning lottery numbers:

FLOWER: Prime numbers. It was all so neat, and elegant. Numbers that refuse to cooperate, that don't change or divide. Numbers that remain the same for all eternity.
STONE: 3, 7, 11, 13, 19, 23, 31, . . .
FLOWER: Ah, it was the magic combination, the key to the gates of heaven.

We're very happy for Stone and Flower. We're just not sure why the meaning of "prime number" doesn't also remain the same for all eternity: what happened to 2, 5, 17, and 29?[1]

Everybody knows that if a positive integer ends in 2 or 5 then it is not prime (unless the number is 2 or 5 itself). Nonetheless, Leaven, the math genius in *Cube* (1997), takes a *very* long time figuring out that 645 and 372 are not prime numbers. She then sees immediately that $649 = 11 \times 59$, and so 649 is not prime. Also, in the same movie, Kazan, an idiot savant, is supposedly very good at factoring numbers, but actually gets quite a few wrong; see chapter 6 ("Escape from the Cube") for details.

A more testing game of "prime not-prime" is played by McKay and Zelenka in the episode "Hot Zone" of *Stargate: Atlantis* (2004). They tease the jockish Lieutenant Ford for refusing to play, and Ford misses a great opportunity to squish them: they have just incorrectly declared that 4021 is not prime.

Finally, returning to *Contact*, Ellie and her colleagues have pinpointed a signal from outer space. Initially, we hear two pulses:

ELLIE: Come on. [We hear three pulses.]
ELLIE: All right. It's restarting. Wait a minute, those are numbers. That was 3, the one before it was 2. Um, base ten numbers, just start counting now and see how far we can get. [Five pulses.]
WILLIE: Five. [Seven pulses.]
ELLIE: Seven. Those are primes 2, 3, 5, 7. Those are all prime numbers, there's no way this is a natural phenomenon!

That may well be. But, as nitpickers, we have to point out that we have no idea what a "base ten number" is.

---

[1] In Paul Auster's excellent novel, upon which the movie is based, the same list of primes is given, except 11 is also omitted. Possibly, Auster and the director just liked the sound of certain primes.

18.5 Slips of the Tongue

In the Z-grade horror movie *Shrieker* (1997), the heroine math major Clark is apparently studying "multidimensional topography." However, unless the Earth suddenly became four-dimensional, we assume she meant "multidimensional topology."

In the episode "Trees Made of Glass, Part 2" (2005) of the TV series *Threshold*, the resident math genius Arthur Ramsey mentions "isomorphic group therapy." Just "group theory" would have made (a little bit) more sense.[2]

18.6 Less Is More

In *Karate Kid, Part II* (1986), the Karate Kid discovers that the bad guys are cheating farmers by using weights that are too light. Actually, since the weights go on one side of the scales and the farmers' goods on the other, this means that the bad guys were cheating themselves.

18.7 Simple Arithmetic?

In *The World Is not Enough* (1999), a bomb is traveling at 70 miles per hour and is 106 miles from its target. James Bond immediately declares that they have 78 minutes to stop it. In fact, they have 91 minutes.

In *Entrapment* (1999), Catherine Zeta-Jones says she wants a security clock to lose 10 seconds. She then rigs the clock to lose $\frac{1}{10}$ of a second each minute, for an hour.

In *Bloodhounds of Broadway* (1952), Numbers Foster is a bookie who practices multiplication to help himself relax. At one point, he declares that $52 \times 95 = 5044$.

In *Butterfly Dreaming* (2008), math professor Rob estimates the number of leaves on a tree to be 2^{12}: "about 8000."

In the episode "From Agnes with Love" of *The Twilight Zone* (1964), Elwood is testing Agnes, his jealous and erratic computer. He asks her to calculate "the first prime number larger than the 17th root of 9,000,355,126,606." Agnes "correctly" gives the answer as 5, when in fact the answer is 7. (Agnes's answer would actually have been correct if Elwood had asked for a number around nine billion instead of nine trillion.)

In "The Mind Robber" story of *Doctor Who* (1968), the Master explains how much he has written:

---

[2] John Stillwell, a writer of many excellent university textbooks, once received the draft cover for his latest book *Topological Group Theory*, which read "Topological Group Therapy." Luckily, this mistake was caught before the book was sent off to the printers.

MASTER: For twenty-five years, I delivered five thousand words every week.
ZOE: Well, that's over half a million words!

Zoe is correct, and the Master has indeed delivered *over* half a million words: six million over.

In *Super Mario Bros.* (1993), a lizard-man exits an evolution machine. He promptly, and incorrectly, declares that the square root of 26,481 is 191. He should have started with 36,481.

In *Il Posto* (1961), the main character applies for a job. A main part of the assessment test is to solve the following problem: *We have a roll of copper wire 520 meters long. Three-quarters of it are cut off. Of the remainder, we cut off four-fifths. How many centimeters of wire are left on the roll?* He's given one hour to solve this problem, and his answer, 24, is wrong. (The correct answer is 26.) Still, he gets the job.

18.8 A Very Tough Quadratic

In *Outside Providence* (1999), the teacher solves the quadratic equation $2X^2 + 7X + 3 = 0$ on the blackboard and then, with the solution still there, asks Dunphy to solve it; Dunphy fails. Later, after lots of hard work, Dunphy has to solve the exact same equation; this time he succeeds.

18.9 The Algebra Problem

In the episode "The Algebra Problem" in the TV series *Meet Corliss Archer* (1954), everybody and their uncle is trying to solve Corliss's "algebra" homework problem:

CORLISS: There are 17 prisoners in a jail, and there are 9 policemen in the courthouse, but they have one police car that holds 6 people, including the driver, and only a policeman can drive it. At no time can there be more prisoners than policemen, in the courthouse, in the jail, or in the car. The policemen have to bring the prisoners to the courthouse. How can all the prisoners be moved in 5 round trips?

The problem is impossible for a number of simple reasons, but no one seems to notice. At the end of the episode, it turns out that the teacher made a mistake: there are 15 prisoners, and 7 trips are permitted. However, the problem is still impossible.

18.10 A Tough Competition

In the episode "Looks and Books" of *Freaks and Geeks* (2000), Lindsay rejoins the mathletes and takes part in a competition. In the longest ever blooper,

every answer in the competition is incorrect: the hour hand of a clock moves
0.4 radians in 48 minutes; a rhombus with long diagonal 10 and large angle
100 degrees has area 42; if arcsin $X = 2$ arccos X then $X = 0.9$; the inscribed
sphere of a cube has 0.52 times the volume of the cube.

18.11 Scary Geometry

In *The Man Without a Face*, (1993) Mel Gibson plays a scary man, teaching
his pupil Norstadt some even scarier geometry; see figure 18.4:

Fig. 18.4 A very unlikely right angle.

SCARY MAN: Draw a circle ABC. *Draw within it any straight line* AB. *Now
bisect* AB *at* D *and draw a straight line* DC *at right angles to* AB. *You are
following so far Norstadt?*
NORSTADT: Yes, Sir.
SCARY MAN: Okay, any other straight line AC. *Bisect* AC *and you get the
center of the circle.*

In *Cabiria* (1914), we see Archimedes designing his parabolic mirrors. Unfortunately, he's using a compass to do so (figure 18.5).

In *Stargate* (1994), Dr. Jackson is explaining a strange astronomical scroll
to General West:

*JACKSON: Now these constellations were placed in a unique order forming
a map or an address of sorts. Seven points to outline a course to a position.
And, uh, to find a destination in any three-dimensional space you need six
points to determine an exact location.*

Fig. 18.5 Archimedes and his famous compass?

WEST: You said you needed seven points.
JACKSON: Well, no, six for the destination. But to chart a course you need a point of origin (figure 18.6).

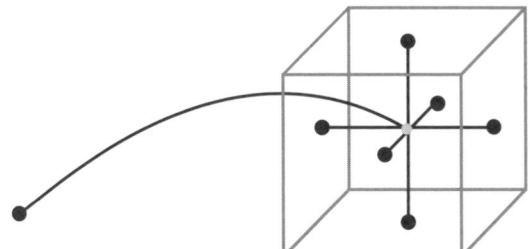

Fig. 18.6 To locate the gray point, you need the six midpoints of faces of the cube and one point outside the cube.

Jackson seems to be confusing points in space with coordinates for points. Given a frame of reference, three coordinates are needed. To establish a frame of reference, four constellations could be chosen, one at the origin and then three along the chosen coordinate axes. Then any point in that coordinate system could be specified with three coordinates: a total of seven pieces of information.

And a final, quick, coordinate geometry blooper: in *Annapolis* (2006), Navy recruits learn calculus and plot their maps with strange bearings: "65 degrees 85 minutes north."

Chapter 19
The Funny Files

The previous chapter was dedicated to listing the funniest mathematical bloopers. Here, we list the funniest math scenes that are *intentionally* funny.

19.1 Sex

Rules of Attraction (2002)

Lauren and Lara are discussing safe sex:

LARA: Okay, we'll do the math. If a condom is 98% safe, and he wears two, then you're 196% safe. A better percentage than the pill can offer.
LAUREN: I don't think it works that way, Lara. Abstinence is 100% safe, which is less a percentage than—
LARA: Whatever. I don't care, I don't major in math.

The Favor (1994)

The mathematician Peter is in the kitchen making a sandwich and rabbiting on about nonlinear dynamics. His wife Kathy walks in, dressed in a sexy nightie, and tries various food-based techniques to entice Peter—totally without success. Finally he concludes she's hungry and offers her a sandwich.

Paradisio (1961)

A math professor comes into possession of glasses that make people appear nude. He somehow uses the area formula πr^2 to estimate distances for his photographs.

Succubus (1968)

A pianist is playing from a school geometry textbook (figure 19.1). This inspires an attractive woman to take off her clothes.

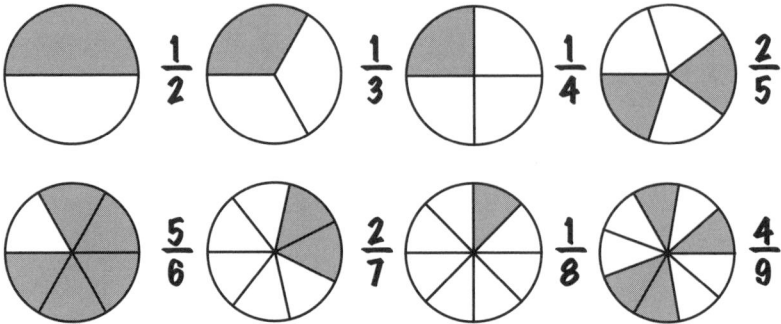

Fig. 19.1 Some of the erotic drawings in *Succubus.*

19.2 Geometry

For more details on the first five entries, see chapter 14 ("Pythagoras and Fermat Go to the Movies").

The Wizard of Oz (1939)

The Scarecrow proves to himself that he has a brain by misquoting Pythagoras's theorem: "The sum of the square roots of any two sides of an isosceles triangle is equal to the square root of the remaining side." Homer Simpson rips off the Scarecrow's Pythagorean declaration in the episode "$pringfield (or, How I Learned to Stop Worrying and Love Legalized Gambling)" (1993) (figure 19.2).

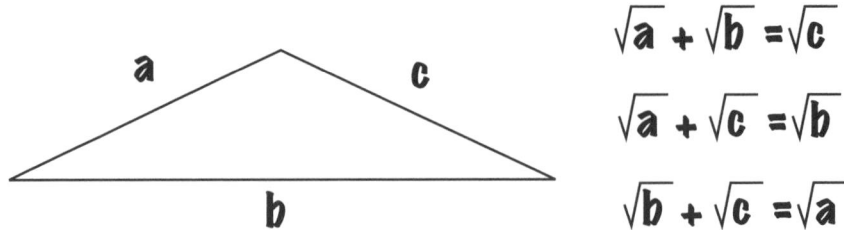

Fig. 19.2 The Scarecrow's (and Homer's) mutilation of Pythagoras's theorem.

Merry Andrew (1958)

Danny Kaye sings and dances Pythagoras's theorem and other miscellaneous bits of geometry.

30 Virgins and Pythagoras (1977)

Czech heartthrob Karel Gott plays a math teacher who becomes a rock star performing the song "Thanks, Mr. Pythagoras."

Fermat's Last Tango (2001)

A wonderful, funny mathematical musical about Fermat's last theorem.

Bedazzled (2000)

Elizabeth Hurley, as a devil schoolteacher, dismisses Fermat's last theorem as useless.

Batman (1966)

BATMAN: What kind of creature would gobble up a bird in a tree?
CHIEF O'HARA: Heaven protect us.
EVERYBODY EXCEPT BATMAN: A cat!
BATMAN: Yes, Gentlemen, the criminal catless to this entire affair: our old arch enemy, Catwoman.
COMMISSIONER GORDON: Penguin, Joker, Riddler, and Catwoman, too. The sum of the angles of that rectangle is too monstrous to contemplate!

Battlefield Earth (2000)

A tedious movie, but it does contain one interesting and funny math scene. After being passed through a learning machine by the monster captors, our hero Tyler explains some mathematics to his fellow prisoners:

PRISONER 1: This is the monster's language?
HERO: No, no, this is mathematics. This is the unifying language of the entire Universe. Look, this symbol is called the triangle. If all its sides are equal, then these three angles must also be equal. [Figure 19.3]
PRISONER 2: Equal to what?
HERO: To each other. It's the basic foundation of Euclidean geometry.
PRISONER 3: Seems pretty hard to understand.

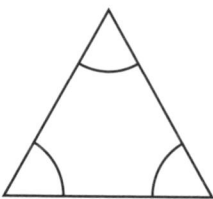

Fig. 19.3 If the sides are equal, then the angles must also be equal. Very tough stuff.

Better Off Dead (1985)

Contains a hilarious scene featuring a math teacher who, despite talking some boring and nonsensical geometry, is treated like a superstar by his students; see chapter 13 ("Beautiful Math, or Better Off Dead").

Buffy the Vampire Slayer: "Tough Love" (2001)

WILLOW: We were acting out a geometry problem—so we made a triangle with our bodies, and that's when I called Xander obtuse, and he got really grumpy. And then Dawn said we were a cute triangle and, well, hilarity ensued.

Family Guy: "When You Wish Upon a Weinstein" (2003)

A fantasy scene demonstrating Chris's need to learn math to function in the real world. He's getting directions at a gas station: "Okay, now whatcha gotta do is go down the road, past the old Johnson place, and you're gonna find two roads, one parallel and one perpendicular. Now, keep going until you come to a highway that bisects it at a 45 degree angle. Solve for X."

19.3 Arithmetic

In the Navy (1941)

Our all-time favorite math scene, in which Abbott and Costello prove that $7 \times 13 = 28$; see chapter 10 ("$7 \times 13 = 28$") for details, and a list of the many other movie versions.

Team America: World Police (2004)

This movie features a very funny running joke:

COMPUTER: Several terrorist groups are being organized for one massive worldwide attack.
SPOTTSWOODE: From what intelligence has gathered it would be nine eleven times a hundred.
GARY: Nine eleven times a hundred. Jesus, that's—
SPOTTSWOODE: Yes, Ninety-one thousand, one hundred . . .

SPOTTSWOODE: Team, if the Durkadurkastanis have weapons of mass destruction, I'm afraid it could be nine-eleven times a thousand.
SARA: Jesus, you mean—
SPOTTSWOODE: Yes, nine hundred and eleven thousand . . .

LISA: So you're the bastard planning nine-eleven times a thousand.
KIM JONG IL: Nooh! You think so small. You see, I'm about to have an elaborate peace ceremony. And while all the World's most important people are distracted here, I will detonate the WMDs, which I have given to terrorists all around the globe. It will be nine-eleven times two thousand three hundred and fifty-six.
GARY: My God, that's—I don't even know what that is!
KIM JONG IL: Nobody does!

Stand and Deliver (1988)

The fingerman scene, in which the hero Escalante shows the reluctant Chuco how to use his fingers to multiply by nine; see chapter 3 ("Escalante Stands and Delivers") for details.

Love and Death (1975)

Boris (Woody Allen) with a drill sergeant:

SERGEANT: One, two, one, two, one, two,—
BORIS: Three comes next, if you're having any trouble.

Later, Boris and Sonja plan their kissing:

SONJA: Kiss me.
BORIS: Which one do you want?
SONJA: Give me number eight.
BORIS: Number eight. That's two fours. That's easy.

The Road to Hong Kong (1962)

Bob Hope and Bing Crosby's last road movie, with a couple of fun mathematical references. For example, to check that Chester (Hope) has regained his memory, Harry (Crosby) quizzes him:

HARRY: How much is two and two?
CHESTER: Four.
HARRY: Four and four?
CHESTER: Eight.
HARRY: Eight and eight?
CHESTER: Seventeen.
HARRY: That's our boy!

Clueless (1995)

Cher (Alicia Silverstone) is giving advice to Elton on how to look after Tai, who has been knocked to the floor:

CHER: It's a concussion. You have to keep her conscious, okay? Ask her questions.
ELTON: What's seven times seven?
CHER: Stuff she knows!

Dick Tracy (1990)

Madonna sings a fun math song: "Count your blessings, one, two, three. I just hate keeping score. Any number is fine with me. As long as it's more. As long as it's more! I'm no mathematician, all I know is addition. I find counting a bore. Keep the number mounting, your accountant does the counting . . ."

Mr. Bean: "Good Night, Mr. Bean" (1990)

To get to sleep Mr. Bean counts sheep in a picture. Losing patience, he uses a calculator to count them as 27×15, and immediately drops off.

Evil Roy Slade (1972)

Betsy is trying to rehabilitate the outlaw Slade:

BETSY: Let's try some arithmetic: If you had six apples and your neighbor took three of them what would you have?
SLADE: A dead neighbor and all six apples.

Straw Dogs (1971)

In one scene, the wife of the mathematician (Dustin Hoffman) decides to annoy him by changing a plus to a minus on his blackboard. It works (figure 19.4). Actually, the formula also changes in mysterious unintended ways; see figure 19.5. A similar scene appears in the movie *Ring* (1998).

Blackadder II: "Head" (1986)

BLACKADDER: Right Baldrick, let's try again. This is called adding. If I have two beans and then I add two more beans, what do I have?
BALDRICK: Some beans.
BLACKADDER: Yes—and no. Let's try again, shall we? I have two beans, then I add two more beans what does that make?
BALDRICK: A very small casserole.
BLACKADDER: Baldrick, the ape creatures of the Indus have mastered this. Now, try again. One, two, three, four! So how many are there?
BALDRICK: Three.
BLACKADDER: What?
BALDRICK: And that one.

Fig. 19.4 Dustin Hoffmann spotting the destruction of his plus sign in *Straw Dogs*.

$$\frac{dP}{d\mu} - \frac{MG}{R^2} + \frac{4\pi RPG}{C^2}$$

Blackboard from far away.
Wife has an idea for a joke.

$$\frac{dP}{d\mu} + \frac{MG}{R^2} + \frac{4\pi RPG}{C^2}$$

Same blackboard close up.

$$\frac{dP}{d\mu} + \frac{MG}{R^2} - \frac{4\pi RPG}{C^2}$$

Blackboard close up after wife
has changed the second plus
to a minus.

$$\frac{dP}{d\mu} - \frac{MG}{R^2} - \frac{4\pi RPG}{C^2}$$

Blackboard from far away.
Husband notices something
is wrong.

$$\frac{dP}{d\mu} + \frac{MG}{R^2} - \frac{4\pi RPG}{C^2}$$

Same blackboard close up.

Fig. 19.5 The *Straw Dogs* mystery: These are consecutive views of the formula on the blackboard. To annoy the mathematician his wife changes the plus sign on the right to a minus sign. Strangely the sign on the left also keeps changing. Since this is not a ghost story, it is presumably not intentional.

BLACKADDER: Three and that one. So if I add that one to the three what will I have?
BALDRICK: Ah! Some beans.

Mozart and the Whale (2005)

This is a lovely movie, a very sweet romance between Donald and Gracie, both afflicted with Asperger's syndrome. Donald is a mathematical prodigy, and there are a number of funny and touching math scenes. At one point Donald is looking to call Gracie and is discussing it with his similarly handicapped friends:

DONALD: Gonna find another phone. I know the number.
GRACIE: Is it as good as a number as say two-eight-oh-nine?
DONALD: Two-eight-oh-nine is 53 squared, and 5 plus 3 is 8. The cube root of 8 is 2, which is her dress size. I bought her this beautiful little dress the other day that has these little flowers on it ...
ROGER: Six!
DONALD: It's not a richly textured number, but it is her shoe size.

Knocked Up (2007)

Ten years ago Ben got $14,000, and he has $900 left. "So that should last me for like—I mean I'm not a mathematician, but like another two years, or some shit, I think." He's out by sixteen months.

Marius (1931)

César teaches Marius how to mix an exotic drink: one very small third of curacao, one third of citron, a large third of picon, and, finally, a really large third of water.

Little Man Tate (1991)

Fred's teacher writes the numbers 1, 2, 3, 4, 5, 6, 7, 8, 9, and 10 on the blackboard:

TEACHER: Who can tell me how many of these numbers are divisible by two? Anybody? Fred!
FRED: Hmm?
TEACHER: I know that you can tell me. How many of these numbers are divisible by two?
FRED: Um, all of them.

Small Time Crooks (2000)

The small time crooks, led by Ray (Woody Allen), are discussing how to split their prospective haul:

DENNY: How much is my share?
RAY: I figure it's got to be two million, right? I'm counting the jewellery in there, too. Divided four ways that's half a million bucks apiece.
RAY: What about Frenchy?
BENNY: What about Frenchy? She's just a front.
RAY: Yeah, but without Frenchy we're dead.

TOMMY: Come on we can get any broad to sell cookies. Geez.
BENNY: Alright, you know what I say? She gets a share, but not a full share.
DENNY: Yeah, what if we each get a fourth, and she gets like a third?
BENNY: What are you nuts? Then she'd be getting more than us?
DENNY: How you figure?
BENNY: Where are you gonna get four fourths and a third? Can't you add?
DENNY: I don't do fractions, all right?

Clue (1985)

Wadsworth the butler and Miss Scarlet are debating whether her gun is empty:

WADSWORTH: The game's up Scarlet. There are no more bullets left in that gun.
MISS SCARLET: Come on, I'm not going to fall for that old trick.
WADSWORTH: It's not a trick. There was one shot at Mr. Body in the study, two for the chandelier, two at the lounge door, and one for the singing telegram.
MISS SCARLET: That's not six.
WADSWORTH: One plus two plus two plus one [1+2+2+1].
MISS SCARLET: Nah-ah. There was only one shot that got the chandelier. That's one plus two plus one plus one [1+2+1+1].
WADSWORTH: Even if you were right, that would be one plus one plus two plus one [1+1+2+1], not one plus two plus one plus one.
MISS SCARLET: Okay, fine one plus two plus one—Shut up! Point is, there is one more bullet left in this gun, and guess who's going to get it?

Life Is Beautiful (1997)

During WWII, the local Nazis in an Italian town are discussing how to save money:

SCHOOLMASTER: And I'm talking about Berlin, the countryside, 3rd grade. Listen to this one. Really, it's quite shocking. The problem: Supporting a lunatic costs the state 4 marks a day, supporting a cripple costs $4\frac{1}{2}$ marks, an epileptic $3\frac{1}{2}$ marks. Figuring the average is 4 marks a day and there are 300,000 of them, how much would the state save if these individuals were simply eliminated?
SHOCKED [for the right reasons] *BYSTANDER: Completely unbelievable!*
SCHOOLMASTER: That's exactly how I reacted. Completely unbelievable. I can't believe an elementary school child is expected to solve something like this. It's a difficult calculation. The proportions, the percentages, they need some algebra to solve these equations, right? That would be high-school material for us.

OTHER NAZI: No, no, it's just multiplication. What did you say it was, 300,000 cripples?
SCHOOLMASTER: Yes.
OTHER NAZI: 300,000 times 4. We'd be saving about 1,200,000 marks a day if we killed them all. It's easy.
SCHOOLMASTER: Exactly, bravo, but you're an adult. In Germany 7-year-olds are given this problem to work out. The most amazing race indeed.

19.4 Algebra and Word Problems

Freaky Friday (2003)

Jamie Lee Curtis is caught in her daughter Anna's (Lindsay Lohan's) body. She's reading a math exam question to herself: "The sum of the areas of the shaded regions above in terms of D is equal to (a) D squared times the sum of pi divided by 4 minus D divided by 2, (b) D squared times the sum of pi cubed divided by D minus 2. Now, what is pi again? 3 point something? Oh this is ridiculous, I've never used pi. Anna's never gonna use pi. Why's it called pi anyway? Okay, focus—Or (c) D cubed minus the sum of pi squared minus—"

Twin Peaks: "1.5" (1990)

The sexy Audrey Horne correctly sums up the situation: "I've been doing some research. In real life, there is no algebra."

Quality Street (1937)

Phoebe (Katharine Hepburn) and Susan are struggling to run their little school:

SUSAN: Phoebe. If a herring and a half costs three ha'pence, how many for elevenpence?
PHOEBE: Eleven.
SUSAN: William Smith says it's fifteen, and he's such a big boy. Do you think I ought to contradict him? May I suggest there are differences of opinion about it? One can't be really sure, Phoebe.
PHOEBE: It is eleven! I once worked it out with real herrings.[1]

---

[1] Regrettably, the movie omits some further dialogue from J. M. Barrie's play. Susan goes on to ask what algebra is: "Is it those three-cornered things?" Phoebe replies: "It is X minus Y equals X plus Y and things like that. And all the time you are saying they are equal, you feel in your heart: why should they be?"

I Went Down (1997)

A hilarious scene, with a kidnap victim tied to a bed, and left holding a remote so he can watch TV. He's flicking through the channels, and just gets to an education channel when he drops the remote. He's stuck watching some very nerdy mathematicians proving the fundamental theorem of algebra.

Idiocracy (2006)

An average Joe from today wakes up in the future and finds himself the most intelligent man on Earth. He is forced to take an IQ test: "If you have one bucket that holds 2 gallons, and another bucket that holds 5 gallons, how many buckets do you have?"

Desk Set (1957)

Richard (Spencer Tracy) tests Bunny (Katharine Hepburn):

RICHARD: A train started out at Grand Central, with seventeen passengers aboard and a crew of nine. At 125th Street, four got off and nine got on. At White Plains, three got off and one got on. At Chappaqua, nine got off and four got on. And at each successive stop thereafter nobody got off, nobody got on, till the train reached its next-to-the-last stop, where five people got off and one got on. Then it reached the terminal.
BUNNY: Well that's easy, eleven passengers and a crew of nine.
RICHARD: Uh, that's not the question.
BUNNY: I'm sorry.
RICHARD: How many people got off at Chappaqua?
BUNNY: Nine.
RICHARD: That's correct!
BUNNY: Yes, I know.

Attack Girls' Swim Team Versus the Undead (2007)

Hilarious, crazy scene, in which the math teacher turns into a zombie while asking a word problem. Then he starts chopping off limbs using his metal rulers.

How Green Was My Valley (1941)

Huw and his father and tutor are puzzling over a word problem: "Now then. The bathtub holds 100 gallons. *A* fills it at 20 gallons a minute, and *B* at the rate of 10 gallons a minute—Now then, *C* is a hole that empties it at 5 gallons a minute. How long, um, to fill the tub?" Huw's mother just laughs: "Who would pour water into a bathtub full of holes?"

Southpark: "Le Petit Tourette" (2007)

Mr. Garrison is attempting to explain the multiplication of negative numbers, continually interrupted by Cartman, who is pretending to be afflicted with Tourette syndrome.

Deep Rising (1998)

VILLAIN: Where are we?
HERO: Right there, middle of nowhere.
VILLAIN: And our final destination?
HERO: Right there, middle of nowhere squared.

Futurama: "Bender's Big Score" (2007)

An incredibly intricate time travel plot in this feature length *Futurama* episode. The Harlem Globetrotters use their "razzle-dazzle globetrotter calculus" (variation of parameters and expansion of the Wronskian) to prove the possibility of paradox-free time travel: "Man, that cube root was a real buzzer beater, Clyde."

Little Big League (1994)

Hilarious scene where the Minnesota Twins have to help their teenage manager solve a word problem; see chapter 17 ("Problem Corner").

Die Hard: With a Vengeance (1995)

Bruce Willis and Samuel Jackson clumsily solving a decanting problem to avoid being blown up; see chapter 9 ("Word Problems for Die Hards").

Family Guy: "Mr. Griffin Goes to Washington" (2001)

Peter imagines his kids learning math on the street:

KID 1: Louis left his house at 2:15, and has to travel the distance of 6.2 miles at the rate of 5 miles per hour. What time will Louis arrive?
KID 2: Depends if he stops to see his ho.
KID 1: That's what we call a variable.

Later, a senator proposes fining the El Dorado Tobacco Company infinity billion dollars; another senator suggests that fining them a real number would be more effective.

19.5 Mathematicians in Action

The Siege of Syracuse (1960)

Archimedes uses his parabolic mirror to set a bathing beauty's clothes on fire.

Monty Python Live at the Hollywood Bowl (1982)

Archimedes again, this time competing in the Germany versus Greece philosophers' football match. Archimedes' Eureka moment is to kick the ball, leading to his team scoring against the German goalkeeper, Leibniz.

The Big Bang Theory: "Pilot" (2007)

SHELDON: There's some poor women who's gonna pin her hopes on my sperm. What if she winds up with a toddler who doesn't know if he should use an integral or a differential to solve for the area under a curve?
LEONARD: I'm sure she'll still love him.
SHELDON: I wouldn't.

Flubber (1997)

Professor Phillip Brainard, played by Robin Williams, stumbles into a life drawing class and begins lecturing about gravity. Great improvisation involving the nude models and a dead pheasant.

Battle of the Worlds (1961)

Professor Benson is a brilliant mathematician, but also the funny-crazy star of the movie. Another highlight is all the math scribbled on flower pots.

Wonder Woman: "The Pluto File" (1976)

Professor Warren uses the harmonic equation and other partial differential equations to create earthquakes. He then struggles to figure out how to stop them, but Wonder Woman comes to the rescue. She declares "integral calculus is always problematical," and then shows the professor the critical substitutions.

That's Adequate (1989)

We are shown a scene of the new movie *Einstein on the Bounty*. Albert Einstein, as the young sea captain of a sailing ship, is talking to his crew, with an active volcano in the background:

EINSTEIN: Whoa! Look at those girls on the beaches. Especially the ones that aren't on fire. You know, there's a lot to learn here. You know one can contemplate the secrets of time, space, matter, energy, all that bullshit. Molecules heating, gases expanding, atoms colliding, matter exploding. It's just fucking brilliant! Wait, wait, now let's see. I was on to something there. E—equals—m—a—no. [The volcano erupts.] *Jesus Christ! I got it! Time! Space! Matter! Energy! E equals m fucking c squared! Yeah!* [At that moment, another ship appears.] *Oh, shit! Look, it's the pirates from the Berlin Academy of Physics. They roam the seas looking for radical physicists with new ideas. Off with their heads!*
[Cut to the fighting.]
PIRATE 1: Fuck relativity!
PIRATE 2: Down with new theory!
[Einstein is crossing swords with a third pirate.]
EINSTEIN: E—equals—m—c—squared—you bad boy!

Insignificance (1985)

This movie contains a very long, absolutely brilliant scene in which Marilyn Monroe demonstrates to Albert Einstein the theory of relativity. The movie also contains a great scene in which Einstein describes to Marilyn the shape of the Universe (as a three-dimensional sphere); see chapter 15 ("Survival in the Fourth Dimension").

Eureka: "Blink" (2006)

Two stereotypically ditzy cheerleaders are chatting as they walk to class: "which basically means certain nonlinear dynamical systems, extremely sensitive to initial conditions ..." "exhibits a phenomenon known as chaos. Come on Courtney, that's a textbook definition. You gotta take specific evaluation ..."

Magik and Rose (1999)

A hilarious New Zealand movie, featuring a square-dancing mathematical prodigy who is into chaos theory.

Everybody Loves Raymond: "Ally's F" (2004)

Ray and Debra go to see Ally's math teacher. He is unsympathetic: "The thing about math is, numbers are constant, they're clear. They're—logical, they're organized. Thirteen-year-olds are—not.—Their home lives and their love lives and their social lives, are not my problem. That is my problem, and the answer is pi!" Ray sums him up: "If X equals lame, that guy is four times X."

Lucky Pierre (1974)

The DVD of this French comedy comes without subtitles. However, you don't have to understand a word to enjoy the hilarious scene where Pierre Richard plays a manic teacher giving a math lesson; see page 145 for a screenshot.

19.6 Doing the Impossible

Lambada (1991)

In one scene the supercool math teacher Blade is making a close to impossible three-cushion billiards shot using a protractor and his knowledge of the "Cartesian coordinate system"; see also chapter 5 ("Nitpicking in Mathmagic Land").

Teresa's Tattoo (1994)

Teresa is a PhD student in applied mathematics. There are some very funny scenes featuring her trying (unsuccessfully) to tell various clueless people what she does. Also, when she is heavily drugged, Teresa recites the quadratic formula.

Graveyard Disturbance (1987)

This is a cheap but fun horror movie, where five teenagers get lost in a cemetery building that loops around in an impossible way. One of them tries to make sense of it by referring to Escher's impossible drawings.

Similar impossible geometry pops up in a number of other movies, always in fun ways. In *The Matrix Revolutions* (2003), Neo gets stuck in a train station, which loops around à la Pacman. In the sitcom world of *Pleasantville* (1998), a geography teacher explains how the town loops back on itself; it's a very funny scene, although the DVD commentary makes clear that the writer-director didn't completely understand what he was doing.

There are also a few scenes based on Escher's drawing *Relativity*, a picture of an impossible staircase in which gravity seems to be pulling in three different directions. These include *The Simpsons* episode "Homer the Great" (1995), in which the beginning couch gag is modeled after this drawing; and the staircase in *Labyrinth* (1986), on which David Bowie (who plays the king of the goblins) walks around. Finally, in *The Avengers* (1998) Uma Thurman runs down an Escherlike never-ending staircase and into a Pacman maze of rooms.

Bill and Ted's Bogus Journey (1991)

Bill and Ted end up in Hell and have a very memorable encounter with Colonel Oats, who ask them to do infinitely many pushups; see page 182.

Twilight Zone: "I of Newton" (1985)

A mathematician carelessly offers to sell his soul to the devil, in exchange for solving a math problem. Luckily he turns out to be smarter than the devil and finds a way out of this dilemma. Good sound practical advice about what to do when the devil calls.

Back to the Future Part III (1990)

Featuring a great four-dimensional idea involving traveling into the future across a bridge that hasn't been built yet. At another point in the movie Doc says: "Clara was one in a million. One in a billion. One in a googolplex."

Monty Python and the Holy Grail (1975)

This movie features a hilarious pseudological argument that culminates in the conclusion that a woman is a witch if she weighs the same as a duck (which turns out to be true).

Flash Gordon (1980)

Doctor Zarkov, Prince Barin, and Princess Aura are fleeing and are attempting to open a locked door:

AURA: They've changed the code!
BARIN [romantically]*: I've changed too, Aura.*
AURA [equally romantically]*: And I've changed, too.*
ZARKOV: Oh it's okay I think I can work it out.

As Barin and Aura talk and look deep into each other's eyes, Zarkov focuses on the problem at hand and succeeds in opening the door.

ZARKOV: Ah, I thought it was one of the prime numbers of the Zeeman series. I haven't changed!

Bingo (1991)

Bingo the dog is helping Chuckie with his homework:

CHUCKIE: What is the square root of nine?
BINGO: Woof! Woof! Woof!

Proof (2005)

This excellent movie features a scene where a rock group composed of mathematicians is performing their song entitled "i": It consists of them just standing there not playing at all.

Bullshot (1983)

Bullshot Crummond is the stereotypical gentleman English hero. He is hunting pigeons with his friend Binky Brancaster, and shoots one without looking:

BINKY: Crummond! How the Dickens did you do that?
CRUMMOND: Simple, Binky. By rapidly calculating the pigeon's angle of elevation in the reflection of your monocle, then subtracting the refractive index of its lens, I positioned myself at a complementary axis and fired. It was no challenge at all.

Later Crummond is dancing with Diz White, whose father has been kidnapped in an attempt to gain his secret formula:

CRUMMOND: Now, can you recall your father's formula?
DIZ: Let's think. It started with a capital N, and then there was a little A, followed by a 3. Ah, then there was a squiggle above a tick, and ah—a hot cross bun sign.

I.Q. (1994)

In a scene reminiscent of *Bullshot*, Einstein and his friends are helping pretend-genius Ed "remember" a formula, without letting anybody else see what they're doing [one crosses his fingers and then indicates an equal sign in the same way]: "X equals—" [another holds up one finger straight and then crosses it with another finger horizontally. He follows this up by using thumb and index fingers of both hands to make a "w"] "$1 + w$" [another holds up an ice cube and has it hovering over a pie] "cubed over pi." See figure 19.6.

Fig. 19.6 Signaling "cubed over pi" in *I.Q.*

Thunderbirds: "Sun Probe" (1965)

Brains is asking the robot Brainman a question of life and death:

BRAINS: Now, Brainman, I want you to calculate the following equation. What is the square root to the power of 29 of the trigonometric amplitude of 87 divided by the quantitative hydraxis of 956 to the power of 77. Do you understand the question?
BRAINMAN: Yes.
BRAINS: Off you go then ...
BRAINMAN: 45,969.

Rosencrantz & Guildenstern Are Dead (1990)

The two main characters toss a coin, getting 157 heads in a row, all the while discussing what it means. Trance, a very lucky character in the sci-fi TV series *Andromeda*, tosses 58 tails in a row in the episode "Attempting Screed."

Red Planet Mars (1952)

Scientists are contacted from Mars, and are trying to figure out how to communicate with the "Martians." The breakthrough comes when the head scientist's son gets the idea of using the digits of the decimal expansion of π (while biting into a piece of pie). Hard to describe, but the movie is a crazy must-see.

19.7 What Are the Odds?

My Little Chickadee (1940)

Very funny one-liner from W. C. Fields about whether poker is a game of chance: "Not the way I play it, no."

Jason X (2001)

A cute android named Kay-Em calculates the probability of the team's survival at 12%. After her friend Tsunaron kisses her, she recalculates the odds at 53%. He suggests they try for 100%.

Torchy Gets Her Man (1938)

Gahagan has a mathematical scheme for betting on the horses: "I make my mind go blank until a number pops into it. Then I multiply the number by the same number, and the last number of the result is the horse I plunk it on." He uses his system to pick horse 7 (because $6 \times 6 = 37$), then 4 ($4 \times 4 = 14$), then 9 ($9 \times 9 = 99$), then 7 ($7 \times 7 = 47$), and then 8 ($8 \times 8 = 68$). He wins.

19.8 Odds And Ends

My Stepmother Is an Alien (1988)

The alien stepmother (Kim Bassinger) is being questioned: "What do you do on vacation?" "Math." "What do you do to have fun?" "Graphs!"

Oh God! Book II (1980)

George Burns as God helps a little girl do her math. It turns out he's very good at it. God also admits to a mistake by having made math too hard.

The School of Rock (2003)

Jack Black, pretending to be a teacher, leads his class in a very funny math song: "So get off your ath and do some math."

Ball of Fire (1941)

A fun movie about a team of stuffy academics being charmed by Barbara Stanwyck. When Isaac Newton is raised in conversation, she refers to herself as just another apple. The mathematician tries to find the common denominator of Stanwyck's syncopated dance moves, and argues from relativity that the signpost ran into the academics' car. Later, the academics argue over whether the correct grammar is "two and two is five" or "two and two are five." They also use Archimedes' mirror idea to thwart the bad guys.

The Simpsons and Futurama

There is a lot of mathematics to be found in various *Simpsons* and *Futurama* episodes. As you would expect, just about all of it is very clever and very funny. For comprehensive lists, see the Simpsons Math website and the math section of the webpage entitled La Indoblable Página de Bender Bending Rodriguez.

Harold & Kumar Escape From Guantanamo Bay (2008)

Kumar is writing a love poem called "The square root of 3," when he is interrupted by Vanessa struggling with her "fucking calculus final." Kumar helps her with a double integral. Later, Kumar reads his poem:

I fear that I will always be a lonely number, like root 3.
A 3 is all that's good and right. Why must my 3 keep out of sight
Beneath a vicious square root sign? I wish instead I were a 9.
For 9 could thwart this evil trick with just some quick arithmetic.
I know I'll never see the Sun as 1.7321.
Such is my reality, a sad irrationality.

When Hark!, what is this I see? Another square root of a three
Has quietly come waltzing by. Together now we multiply
To form a number we prefer, rejoicing as an integer.
We break free from our mortal bonds, and with a wave of magic wands
Our square root signs become unglued, and love for me has been renewed.

Part III
Lists

Chapter 20
People Lists

For this chapter, we have compiled a number of annotated lists of mathematical people:

1. Real mathematicians
2. Female mathematicians
3. Notable math teachers and classroom scenes
4. Wizkids
5. Murderous mathematicians
6. Famous actors being mathematical
7. Math consultants

For a comprehensive list of titles and summaries of the relevant movies, check out our MathsMasters website (`www.qedcat.com`). We'll continue to add to that website after this book has appeared.

20.1 Real Mathematicians

Real mathematicians seldom appear as characters in movies, and it seems that many of the mathematical greats have never appeared. The following is essentially a complete list.[1]

Jean d'Alembert appears in *Si Versailles M'Était Conté* (1954), but he has no speaking lines.

Archimedes of Syracuse is one of the most frequently featured mathematicians. He appears in the earliest math movie of which we are aware, *Cabiria* (1914), in which he performs the usual geometrical construction to burn the usual Roman warships. Incredibly, Archimedes is also the hero of a sword and sandals flick, *The Siege of Syracuse* (1960), complete with a final climactic fight (which he wins!). In this movie, Archimedes burns not only Roman

---

[1] We have not included Albert Einstein in the list, since he was much more a physicist than a mathematician. And, needless to say, the appearances of this iconic character are too numerous to list.

warships, but more interestingly the clothing of a bathing beauty. Among Archimedes' bit roles are his being killed by Roman soldiers in *Quest of the Delta Knights* (1993) and his Eureka scene (with towel) in *Gulliver's Travels* (1996). A much funnier role occurs in *Monty Python Live at the Hollywood Bowl* (1982), in which Archimedes stars for Greece against Germany in the philosophers' football match: his Eureka moment comes when he decides to kick the ball.

Charles Babbage collaborates with Ada Byron, and his analytical engine features prominently in *Conceiving Ada* (1997).

Benjamin Banneker, a black American mathematician (more astronomer and engineer) from the eighteenth century, is the main character in the TV movie *Freedom Man* (1989). The movie ends with a very moving voiceover: "My whole life has been a mathematical proof of what a black man can do. I've been a clockmaker, a farmer, an astronomer, a surveyor, an almanac writer, a mathematician.—And I add this, and this. And I hope the sum proves something to the World."

Isaac Beeckman, the seventeenth-century Dutch mathematician, appears posting a problem on gravitation in *Cartesius* (1974).

Ada Byron works together with Charles Babbage on the analytical engine in *Conceiving Ada* (1997).

Renato Caccioppoli, one of the founders of geometric measure theory, commits suicide in *Death of a Neapolitan Mathematician* (1992).

Chen Jingrun, the famous Chinese mathematician who worked on the Goldbach conjecture, is the subject of the nine-hour soapie *Chen Jingrun* (2001).

René Descartes is the subject of Roberto Rossellini's *Cartesius* (1974). The focus is much more on philosophy than mathematics. One extended scene has Descartes solving a gravitational problem posed by the mathematician Isaac Beeckman. Descartes also makes a brief appearance in Rossellini's *Blaise Pascal* (1972) (figure 20.1).

Euclid is haunting the AfterMath in the mathematical musical *Fermat's Last Tango* (2001).

Pierre de Fermat is the vain antihero in the mathematical musical *Fermat's Last Tango* (2001).

Évariste Galois writes down his mathematical discoveries the night before the duel in *Évariste Galois* (1965). The last days of Galois are also the subject of an extended animation sequence in *3:19* (2008).

Carl Friedrich Gauss haunts the AfterMath in *Fermat's Last Tango* (2001).

Kurt Gödel makes an appearance as Albert Einstein's friend in the comedy *I.Q.* (1994).

Fig. 20.1 Marin Mersenne and Blaise Pascal meet with René Decàrtes in *Blaise Pascal*.

Hua Luogeng, the famous Chinese mathematician, appears in *Chen Jingrun* (2001).

Hypatia, the first notable female mathematician in history, is the central character of *Agora* (2009).

Ted Kaczynski, the Harvard-trained mathematician and notorious Unabomber, is the subject of *Unabomber: The True Story* (1996). There is no clear reference to his being a mathematician.

Omar Khayyam, the famous Persian mathematician and poet, appears in a number of movies, although usually much more in the role of a poet. An improbable example is Vincent Price playing the role in *Son of Sinbad* (1955). There are more mathematical references in *The Keeper: The Legend of Omar Khayyam* (2005) and *Omar Khayyam* (1957), the latter with Cornel Wilde in the starring role.

Sonya Kovalevskaya is the subject of the dark *A Hill on the Dark Side of the Moon* (1983).

Gottfried Leibniz is the vanquished goalkeeper in the philosophers' football match in *Monty Python Live at the Hollywood Bowl* (1982).

Marin Mersenne makes brief appearances in *Blaise Pascal* (1972) and *Cartesius* (1974).

John Milnor is the model for Hansen, John Nash's friend and competitor in *A Beautiful Mind* (2001).

Magnus Gustaf Mittag-Leffler makes an appearance in *A Hill on the Dark Side of the Moon* (1983), about the mathematician Sonya Kovalevskaya.

John Nash is the subject of *A Beautiful Mind* (2001).

John von Neumann makes a fleeting appearance in *Race for the Bomb* (1987) (seemingly edited down from a larger role). Also Helinger, John Nash's supervisor in *A Beautiful Mind* (2001), is modeled after von Neumann.

Isaac Newton has brief appearances in many movies. Most famously, he was played by an aging but still funny Harpo Marx, being hit on the head with an apple in *The Story of Mankind* (1957). In a similar vein, Sir Isaac appears in a Mr. Peabody episode of the cartoon *Rocky and Bullwinkle* (1961): he's having trouble proving the law of gravity, because his apples (and banana) keep going up rather than down.

Newton also discusses his *Principia* with Peter the Great in *Peter the Great* (1986), is the Royal Astronomer in *The King's Thief* (1955), and is an alchemist in *Quest of the Delta Knights* (1986). And Newton appears in hologram form in two *Star Trek* episodes: "Descent, Part 1" from *Star Trek: The Next Generation* (1993), and "Deathwish" from *Star Trek: Voyager* (1996).[2]

Newton has a larger role in the AfterMath in *Fermat's Last Tango* (2001). He is also the subject of the romanticized biography *Newton: a Tale of Two Isaacs* (1997). This latter movie is designed for adolescents. It is cheesy but fun, with a little mathematics of gravitation.

Blaise Pascal is the subject of Roberto Rossellini's *Blaise Pascal* (1972), and appears briefly in *Cartesius* (1974).

Pythagoras meets Donald Duck in *Donald in Mathmagic Land* (1959) and haunts the AfterMath in the musical *Fermat's Last Tango* (2001).

Bertrand Russell visits the barber in *Wittgenstein* (1993) (a sly reference to Russell's version of the liar paradox), and Russell also has a minor role in *Tom & Viv* (1994).

Theon of Alexandria, the father of Hypatia, makes an appearance in *Agora* (2009).

Alan Turing is beautifully portrayed in *Breaking the Code* (1996). Thomas Jericho, the main character in *Enigma* (2001), is also modeled on Turing.

Karl Weierstrass makes an appearance in *A Hill on the Dark Side of the Moon* (1983), about the mathematician Sonya Kovalevskaya.

---

[2] Newton is also listed as a character in "The Darking" from *Star Trek: Voyager* (1997). This episode includes a crowd scene of hologram people, but Newton himself does not seem to appear on camera.

Andrew Wiles is one of the main characters (as Daniel Keane) in the mathematical musical *Fermat's Last Tango* (2001).

Nikolai Zhukovsky, the Russian fluid dynamicist, is the subject of the movie *Zhukovsky* (1950).

20.2 Female Mathematicians

Female mathematicians, either real or fictional, are definitely in the minority in the movies. Here's our list.

Agora (2009): The subject of this movie is Hypatia, generally considered to be the first notable female mathematician. Not actually a lot of math.

Antonia's Line (1995): A movie about a wunderkind who becomes a mathematics professor. We see her lecturing on homological algebra.

Clan of the Cave Bear (1986): Daryl Hannah as the first ever mathematician? She is a Cro-Magnon woman, who learns how to count in fives, using a combination of fingers, doubling, and notches on a stick.

Conceiving Ada (1997): The subject of this movie is Ada Byron, the famous mathematician who collaborated with Charles Babbage on his analytical engine.

Cube 2: Hypercube (2002): One of the victims, Mrs. Paley, is an annoying, senile mathematician. She identifies a diagram as a picture of a tesseract (a four-dimensional cube). See chapter 15.

Dangerous Sex Date (2001): A movie consisting of 89 minutes of murder and kinky sex, and 2 seconds of Riemann Christoffel symbols. This seems to identify the main character, a woman, as a mathematician, although she is usually listed as a librarian.

The Giant Claw (1957): Sally Caldwell is "Mademoiselle Mathematician," helping to defeat a murderous chicken from outer space.

A Hill on the Dark Side of the Moon (1983): The subject of this dark movie is the Russian mathematician Sonya Kovalevskaya.

I.Q. (1994): Meg Ryan plays Albert Einstein's niece. Her character is also an accomplished mathematician and physicist.

Inspector Lewis (2006): In this TV pilot, a female student thinks she has found a proof for the Goldbach conjecture, invalidating the work of her professor, which earned him the Fields medal. Although her proof is not correct, the professor realizes that it pinpoints a serious mistake in his work. He murders her to maintain his secret.

It's My Turn (1980): A light romance, with the main character a mathematics professor specializing in finite group theory. She gives a full proof of the snake lemma in the first scene and also demonstrates how to deal with annoying, uppity students. See chapter 12.

Julie Johnson (2001) is a downtrodden white trash housewife, who turns out to be brilliant at mathematics. There is a funny scene where she tries to explain the Cantor set to her daughter (Mischa Barton).

Las Vegas Shakedown (1955): A math teacher travels to Las Vegas to do research for her book, which is supposed to show that the casinos always win (well, duh).

My Teacher's Wife (1995): The smart and attractive wife of the math teacher herself has a PhD in math. She demonstrates that she is very good at mental calculations and calculus.

Presumed Innocent (1990): A murder mystery, and a morality tale for PhD students. The wife of the lead character has been working for ten frustrating years on her math thesis. In the end it turns out that she committed the murder. However, she gets away with it, finishes her PhD, and lives happily ever after.

Proof (2005): Gwyneth Paltrow plays a troubled mathematician, coming to terms with the death of her troubled and famous mathematician father, in whose shadow she has been living. Toward the end of the movie it turns out that she came up with her own major piece of research, outshining the work of her father.

She Wrote the Book (1946): A woman math professor lectures about the radius of convergence of power series. In another very funny scene she discusses whether she likes bridges and applies (correctly!) the law of tangents.

Shrieker (1998): One of the main characters is Clark, a female math student specializing in "multidimensional topography."

Smilla's Sense of Snow (1997): Smilla does mathematical research on snow and ice. At one point she says to a child: "A point is that which cannot be divided. A line is a length without a breadth. This can't possibly interest you."

Spaceways (1953): The crew of the rocket ship includes Dr. Lisa Frank, a "higher mathematician" who "has everything, including emotions, neatly reduced to equations and theorems."

Teresa's Tattoo (1994): Teresa is doing a PhD in applied math. In a number of funny scenes, she tries and fails to communicate her work to nonmathematicians.

20.3 Interesting Math Teachers and Classroom Scenes

There are a many, many movies featuring math teachers. Here is a list of movies with teachers or classroom scenes that are particularly notable.

Au Revoir Les Enfants (1987): A teacher discusses the fact that for a quadrilateral circumscribed around a circle, the sum of two opposite sides equals the sum of the remaining sides (figure 20.2).

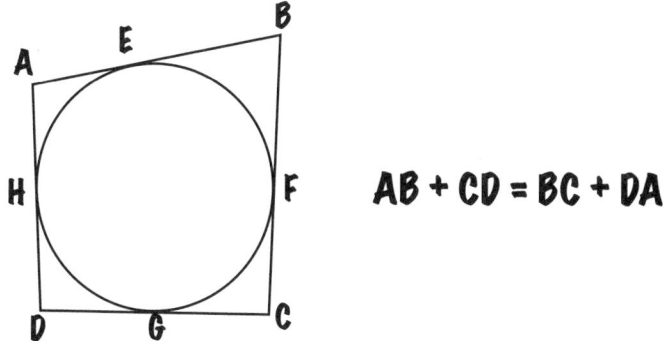

Fig. 20.2 The sum of two opposite sides equals the sum of the remaining sides.

Bedazzled (2000): Elizabeth Hurley plays a devil-teacher, denouncing Fermat's last theorem as useless.

Better Off Dead (1985): A comedy in which a math teacher doing nonsensical geometry is worshipped by his students.

Brutal (2007): A mild-mannered science teacher uses flowers and Fibonacci numbers to plan his murders.

Carry On Teacher (1959): A smug supervisor interrupts a math class to test how kids react to him noting down two-digit numbers with the two digits reversed. He's outsmarted by a kid asking him to write down 33.

Diabolique (1955): One of the main characters is a math teacher. It includes a scene dedicated to a boy finding the area of a regular hexagon in terms of the radius of the circumcircle.

The Infinite Worlds of H. G. Wells (2001): The Pyecraft episode in this miniseries is a fun fantasy about a gentle mathematician working as a math teacher.

Lambada (1990): *Stand and Deliver* with dirty dancing. Blade is a supercool math teacher who inspires his students. Most memorable cheesy scene: Blade demonstrates the usefulness of math by making a near impossible three-cushion pool shot using "the rectangular coordinate system" and a protractor.

Little Man Tate (1991): A movie featuring two mathematical prodigies. In one classroom scene a math teacher writes the numbers 1, 2, 3, 4, 5, 6, 7, 8, 9, and 10 on the blackboard and asks: "Who can tell me how many of these numbers are divisible by two?" One of the prodigies offhandedly answers: "Um, all of them."

Lucky Pierre (1974): The DVD of this French comedy comes without subtitles. But you don't have to understand a word to enjoy the hilarious scene where Pierre Richard plays a manic teacher giving a math lesson.

The Man Without a Face (1993): Mel Gibson as a scary math teacher quizzes a boy on Pythagoras's theorem, and teaches him a Euclidean theorem on how to find the center of a circle: he goofs it!

Mean Girls (2004): A great movie for math spotting on blackboards. The main character is a very pretty girl (Lindsay Lohan) who is also very good at math. Encouraged by her teacher and against the advice of her friends ("social suicide"), she joins the school math club and wins the big math competition for them.

Merry Andrew (1958): Danny Kaye plays a math teacher who teaches Pythagoras's theorem by singing and dancing.

My Teacher's Wife (1995): The math teacher in question is nicely evil, and his smart and pretty wife has a PhD in math.

October Sky (1999): A kid in a coal-mining community wants to become a rocket scientist. He realizes that he'll have to get good at math and is helped by his math teacher.

The Prince and the Pauper (1990): Mickey Mouse as the prince has to sit through a very boring trigonometry lesson.

Rushmore (1998): The movie begins with a dream sequence in which a mathematical underachiever fantasizes that he becomes the class hero, by solving a superdifficult math problem. The problem shown is to calculate the area of an ellipse, which is actually pretty easy.

The School of Rock (2003): Jack Black is intent upon teaching his students rock music. When surprised by the principal, he improvises a very funny math song.

Six Feet Under: "Brotherhood" (2001): This episode includes a class scene, culminating in the math teacher's head exploding.

Stand and Deliver (1988): A terrific movie about the legendary math teacher Jaime Escalante motivating his students in a poor school in East Los Angeles. See chapter 3.

The Virgin Suicides (1999): We see some teaching of Venn diagrams. One of the blackboards is adorned with a messed-up decimal expansion of π. In another scene we see a blackboard with problems related to finding the derivative of a quadratic function.

Willy Wonka & the Chocolate Factory (1971): Some funny scenes involving percentages. One has a teacher using Wonka Bars to teach percentages. Another has a total of 105%. A computer is programmed to find golden tickets based on the "computonian law of probability."

Wish Upon a Star (1996) features a funny and sarcastic math teacher.

The Wonder Years (1988–1993): Season 3 features a few episodes with Mr. Collins, a seemingly cold but very dedicated math teacher. A truly great character.

20.4 Wizkids

Antonia's Line (1995): A movie about a wunderkind who becomes a mathematician. There are wunderkind calculations, and a homological algebra lecture with commutative diagrams on the blackboards.

Dear Brigitte (1965): People want to use a young boy's mathematical abilities to gamble on the horses, but all he wants is to meet Brigitte Bardot. He gets his wish.

Eustice Solves a Problem (2004): A terrific short film, set around a children's quiz program in 1958. Eustice is a little boy who knows all the answers.

Good Will Hunting (1997): Movie about a math prodigy with tons of mathematics. See chapter 1.

Little Man Tate (1991): A movie featuring two mathematical prodigies. Lots of math: math competitions, building tensegrity icosahedra (figure 20.3), and so on.

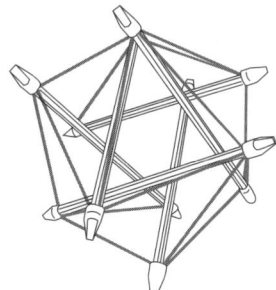

Fig. 20.3 Fred builds himself a tensegrity icosahedron in *Little Man Tate*.

Love, Math and Sex (1997): The heroine is a mathematical prodigy. The nonsex parts include quadratic equations, probability, topology, the Möbius strip, and a math competition.

Matilda (1996): Little Matilda is a math genius and has supernatural abilities. Some very funny scenes.

Mercury Rising (1993): features an autistic boy who is a wiz at cracking ciphers. There are other movies that feature autistic people with strong mathematical abilities in the genius category: *Rain Man* (1988), *Cube* (1997), *Mozart and the Whale* (2005), and *The X Files: "Roland"* (1994).

Raising Genius (2004): This movie contains the most annoying math character of all time. The monumentally whiny Hal is a prodigy who has locked himself in the bathroom to do his math (partial differential equations). And to ogle Lacy, the practicing cheerleader over the fence (Danica McKellar[3]). He refuses to come out ("I can't leave my Lipschitz conditions") and gets annoyed at Lacy's repeated losing the beat on "seven." At the end, he realizes that the "Lacy perturbation" is the key to solving his equations.

Real Genius (1985): A movie about a child prodigy. The running joke in the movie involves three consecutive math lectures in which more and more students are replaced by recording devices. In the last scene even the lecturer is replaced by a recording.

Rushmore (1998): The movie begins with a dream sequence in which a mathematical underachiever fantasizes that he becomes the hero of the class, by solving a superdifficult math problem. The problem shown is to calculate the area of an ellipse, which is quite routine.

Star Trek: The Next Generation (1987–1994): Wesley Crusher, the most annoying character in *Star Trek*, definitely fits the description of a wizkid. For example, in the episode "The Vengeance Factor" (1989), he does homework which involves the "locally Euclidean metrization of a k-fold contravariant Riemannian tensor field."

Village of the Damned (1960): Creepy blond children learn things quickly, including some relativity equations.

Young Einstein (1988): A quirky comedy, where Albert Einstein grows up on an apple farm in Tasmania and rediscovers Newton's theories. He later moves to a great city, falls in love with Marie Curie, and discovers everything else that the (real) Albert Einstein is famous for.

20.5 Mathematicians and Murder

So you're one of those stereotypical, highly strung mathematicians, and something or somebody has just sent you over the edge. How to get even with this special someone and the world in general? If you're looking for some clues in the movies you'll be disappointed. There is no shortage of murderous

---

[3] As well as being an aspiring actress, Danica McKellar is an aspiring mathematician. Her popularizations of mathematics have been well received.

mathematicians, but despite being smart they don't seem to go about their murderous business in particularly smart ways. In fact, all but one in our list get caught.

The Adding Machine (1969): A counting-obsessed accountant named Zero commits a murder.

Bianca (1984): The movie features a murderous math teacher, who also doesn't know his magic squares.

Brutal (2007): A movie about a mild-mannered science teacher who uses flowers and Fibonacci numbers to plan his gruesome murders (figure 20.4). An awful movie, but the main math scene is pure gold.

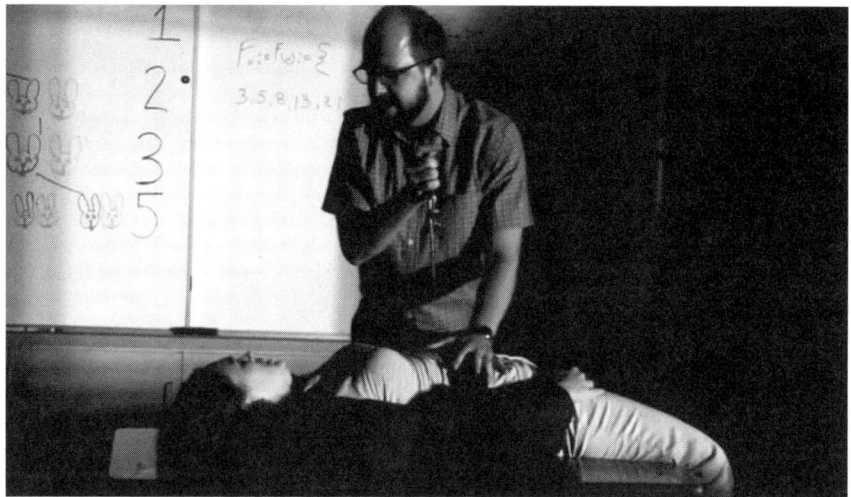

Fig. 20.4 The Fibonacci murderer is about to strike again in *Brutal*.

Casino Royale (2006): M identifies the villain Le Chiffre (whose name roughly translates as "digit") as a mathematical prodigy.

City Homicide: "Hot House" (2010): Some disgruntled ex-students arrange to kill some mathematicians. The bodies are covered by equations. See chapter 8.

Class Action (1991): An actuarial bean counter shows why it is cheaper to deal with lawsuits than to fix a deadly fault in some cars.

Colossus: The Forben Project (1970): Colossus is a supercomputer placed in charge of running America. It uses mathematics to establish communication with its Russian counterpart and then decides to take over the world, murdering a few people in the process.

Death of a Cyclist (1955): The cyclist dies after being run over by a math professor. The movie features cycloids as well as cyclists.

Diabolique (1996): Sharon Stone, as a murderous math teacher, is doing simple algebra in class and is trying to drive her victim mad.

Der Elefant: "Liebesbrief eines Toten" (2004): A math student was murdered 20 years ago, and a fellow student, now professor, is a suspect. One of the detectives apparently understands the math: "The arithmetic mean of omega plus 1.23 to the power of phi equals the limit of the inverse function lambda and the square root of k, whereby k must be a constant. Simple." "Sure. For someone who never had any friends in school."

Fermat's Room (2007): A mathematician who succeeded in proving the Goldbach conjecture concocts an evil plan to prevent another mathematician from publishing first. The centerpiece of this plan is a deadly trap consisting of a shrinking room and four mathematician captives. See chapter 7.

Furuhata Ninzaburō: "Murder of a Mathematician" (1995): Two mathematicians have just received the prestigious Australian Arbuckle award (for mathematicians under 40), for their work on dynamical systems in four-dimensional manifolds. When one is killed, the Columbo-like Furuhata solves the mystery: the other mathematician has killed him to take credit for the solution of Fermat's last theorem. Throughout the episode, Furuhata and the murderer play a simple Nim-like game: Furuhata shows his mastery at the end, by winning and describing the mod 4 calculation used to do so.

Homage (1995): A brilliant young mathematician, with a job offer from Princeton, goes berserk when a young actress rejects him.

Inspector Lewis (2006): A TV pilot featuring the Goldbach conjecture and perfect numbers. A mathematician kills some students to prevent them from publishing a flaw in a proof of a theorem that earned him the Fields Medal.

The Oxford Murders (2008): A fun mystery with two mathematicians taking on the roles of detectives (figure 20.5).

Fig. 20.5 A mysterious pattern is an important clue in *Oxford Murders*.

Presumed Innocent (1990): A jealous wife (who happens to be a mathematician working on her dissertation) commits murder, and gets away with it.

The Seven-Per-Cent Solution (1976): The image of Sherlock Holmes' nemesis Moriarty as a murderous mathematician is challenged, when Sherlock Holmes declares: "The only time Professor Moriarty really occupied the role of my nemesis was when it took him three weeks to make clear to me the mysteries of elementary calculus."

Straw Dogs (1971): Dustin Hoffmann as a mathematician driven to extremes to defend his house and family. Beforehand, he counts to 100 in binary, and there are some good calculus blackboards.

Suspect X (2008): A terrific battle of wits between a physicist-detective and a mathematician-suspect. "But his specialty is math, not homicide." "Yes, but homicide would prove an easier challenge." A stunning scene with the mathematician in a prison cell, as he imagines the four color theorem filling the ceiling of the cell.

Suicide

Believers (2007): A movie about a math cult whose leader has discovered a formula that will transport its members to a safe place before the end of the world. The cult members have the formula tattooed on their bodies and all commit suicide (before the end of the world).

Death of a Neapolitan Mathematician (1992): The story of the Italian mathematician Renato Caccioppoli that ends with his suicide. Some nice calculus scenes and interesting insights into the world of mathematicians.

Imperativ (1982): A dark movie about a suicidal math professor. He specializes in probability theory, and one of his students works out a system to win at roulette. However, the professor is more interested in Russian roulette.

Parallel Life (2010): A crazy mathematician believes (correctly as it turns out) that his life parallels Kurt Gödel's. He writes bizarre probabilistic mathematics on the walls, screams "A coincidence is mathematically impossible," and eventually starves himself to death (as did Gödel).

Nobody Likes Mathematicians

Serial Mom (1994): A math teacher is murdered by a homicidal mother.

A Great Opportunity Missed

The Bishop Murder Case is a Philo Vance mystery, where all the victims and all the suspects are mathematicians. Sadly, in the 1930 movie version all the characters have been turned into chess players: there's not a scrap of mathematics.

20.6 Famous Actors Being Mathematical

Abbott and Costello can be admired in serious mathematical comedy action in *In the Navy* (1941), *Buck Privates* (1941), and many others.

Woody Allen in *Love and Death* (1975) and *Small Time Crooks* (2000) has a little fun with math.

John Astin knows his math as *Evil Roy Slade* (1972). His response to the question "Let's try some arithmetic: If you had six apples and your neighbor took three of them what would you have?" "A dead neighbor and all six apples."

Rowan Atkinson has a few stabs at math as Mr. Bean. In episode 1.1 (1990), Mr. Bean's first ever skit, he flails helplessly in a math exam full of calculus, having instead practiced trigonometry. With two minutes to go, he discovers the alternative trigonometry exam. In the episode "Good Night, Mr. Bean" (1995) he counts sheep in a picture. Losing patience, he uses a calculator to count them as 27×15, and immediately drops off. In *Blackadder II:* "Head" (1986) he attempts and fails to teach Baldrick how to add 2 and 2.

Drew Barrymore joins a math club in *Never Been Kissed* (1999) and takes part in math competitions.

Mischa Barton as the daughter in *Julie Johnson* (2001) has the Cantor set explained to her, and is less than impressed.

Kim Basinger as the alien stepmother in *My Stepmother Is an Alien* (1988) does (very little) math, and graphs for fun.

Jack Black improvises a very funny math song in *The School of Rock* (2003).

David Bowie is the king of the goblins in *Labyrinth* (1986), having fun posing the liar paradox and running around in a maze reminiscent of Escher's drawing *Relativity*.

Jeff Bridges, as the unworldly mathematician in *The Mirror Has Two Faces* (1996) likes prime numbers.

George Burns, as God in *Oh God! Book II* (1980), helps a little girl do her math. He also admits to a mistake, having made math too hard.

Tia Carrere has a PhD in mathematics and proves that she knows her calculus in *My Teacher's Wife* (1995).

Jim Carrey obsesses about the number 23 in *The Number 23* (2007).

George Clooney "does the math" and "imagines the odds" in *Ocean's Eleven* (2001). In *Up in the Air* (2009), Clooney calculates the time a colleague loses by checking in luggage: "35 minutes a flight. I travel 270 days a year. That's 157 hours. That makes 7 days." Later, he's quizzed about his goal of accumulating ten million flier miles: "Isn't ten million just a number?" "Pi is just a number."

Jennifer Connelly solves the liar paradox as the main character in *Labyrinth* (1986) and studies mathematics as Alicia Nash in *A Beautiful Mind* (2001).

Gary Cooper is stuck behind a merry-go-round in *Cloak and Dagger* (1946). To pass the time, he calculates the line integral of a sine curve describing the path of one of the horses. In the reconstruction shown in figure 20.6, r stands for the radius of the merry-go-round and a for the amplitude of one of the horses' movements. See whether you can spot his mistakes. In *Ball of Fire* (1941), he discusses the grammar of "two plus two is/are five."

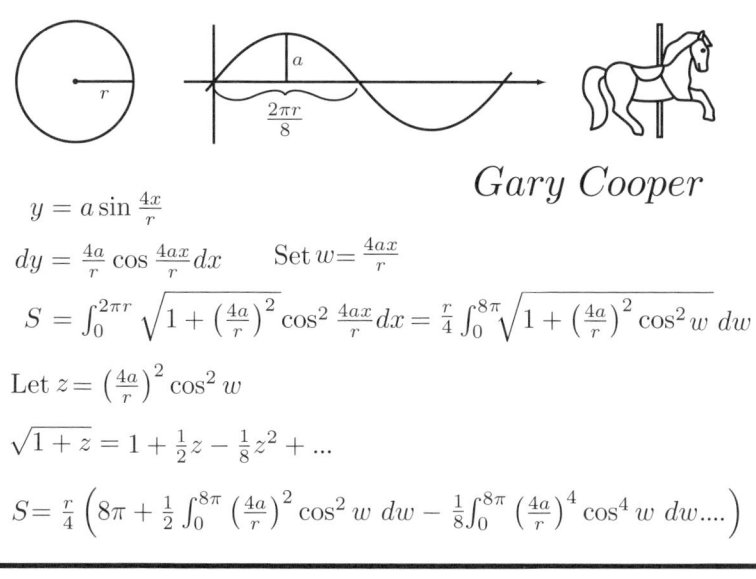

Gary Cooper

$$y = a \sin \tfrac{4x}{r}$$

$$dy = \tfrac{4a}{r} \cos \tfrac{4ax}{r} dx \qquad \text{Set } w = \tfrac{4ax}{r}$$

$$S = \int_0^{2\pi r} \sqrt{1 + \left(\tfrac{4a}{r}\right)^2 \cos^2 \tfrac{4ax}{r}} \, dx = \tfrac{r}{4} \int_0^{8\pi} \sqrt{1 + \left(\tfrac{4a}{r}\right)^2 \cos^2 w} \, dw$$

Let $z = \left(\tfrac{4a}{r}\right)^2 \cos^2 w$

$$\sqrt{1+z} = 1 + \tfrac{1}{2} z - \tfrac{1}{8} z^2 + \dots$$

$$S = \tfrac{r}{4} \left(8\pi + \tfrac{1}{2} \int_0^{8\pi} \left(\tfrac{4a}{r}\right)^2 \cos^2 w \, dw - \tfrac{1}{8} \int_0^{8\pi} \left(\tfrac{4a}{r}\right)^4 \cos^4 w \, dw \dots \right)$$

Isaac Newton

$$y = a \sin \tfrac{4x}{r}$$

$$dy = \tfrac{4a}{r} \cos \tfrac{4x}{r} dx \qquad \text{Set } w = \tfrac{4x}{r}$$

$$S = \int_0^{2\pi r} \sqrt{1 + \left(\tfrac{4a}{r}\right)^2 \cos^2 \tfrac{4x}{r}} \, dx = \tfrac{r}{4} \int_0^{8\pi} \sqrt{1 + \left(\tfrac{4a}{r}\right)^2 \cos^2 w} \, dw$$

Let $z = \left(\tfrac{4a}{r}\right)^2 \cos^2 w$

$$\sqrt{1+z} = 1 + \tfrac{1}{2} z - \tfrac{1}{8} z^2 + \dots$$

$$S = \tfrac{r}{4} \left(8\pi + \tfrac{1}{2} \int_0^{8\pi} \left(\tfrac{4a}{r}\right)^2 \cos^2 w \, dw - \tfrac{1}{8} \int_0^{8\pi} \left(\tfrac{4a}{r}\right)^4 \cos^4 w \, dw \dots \right)$$

Fig. 20.6 Calculations in *Cloak and Dagger*, and what Cooper should have done.

Daniel Craig as Werner Heisenberg in *Copenhagen* (2002) talks about probability and the beauty of math.

Bing Crosby and **Bob Hope** have a couple of very funny math routines in *The Road to Hong Kong* (1962).

Russell Crowe plays the mathematical genius John Nash in *A Beautiful Mind* (2001).

Jamie Lee Curtis is caught in her daughter Anna's (Lindsay Lohan's) body in *Freaky Friday* (2003) and at one point is trying to remember π. "Now, what is pi again? 3 point something? Oh this is ridiculous, I've never used pi. Anna's never gonna use pi. Why's it called pi anyway? Okay, focus—"

Matt Damon is the mathematical wunderkind in *Good Will Hunting* (1997).

Ossie Davis is the mathematician Benjamin Banneker in *Freedom Man* (1989).

Gérard Depardieu is a writer who loves math in *A Pure Formality* (1994). He even knows about points at infinity.

Henry Fonda, who is a member of a jury, uses probability to make his point in *12 Angry Men* (1957).

Jodie Foster is a SETI scientist in prime number action in *Contact* (1997) and is the mother of the mathematical wunderkind in *Little Man Tate* (1991).

Brendan Fraser as *George of the Jungle* (1997) calculates the angle and velocity of his swing before swinging to the rescue of a parachutist entangled in the cables of the Golden Gate Bridge.

Mel Gibson plays a very scary-looking math teacher in *The Man Without a Face* (1993).

Jeff Goldblum, as the mathematician in *Jurassic Park* (1993) and *Jurassic Park II (The Lost World)* (1997) uses chaos theory to impress a woman.

Tom Hanks helps a boy with his algebra homework in *Big* (1988), knows his math when he gets stranded on a deserted island in *Cast Away* (2000), and confronts the Fibonacci numbers in *The Da Vinci Code* (2006).

Daryl Hannah is the first ever mathematician in *Clan of the Cave Bear* (1986).

Sherman Hemsley in *The Twilight Zone:* "I of Newton" (1985) exclaims as a frustrated mathematician: "I'd sell my soul to get this right." The devil promptly appears to accept the trade.

Dustin Hoffmann plays a rampaging mathematician in *Straw Dogs* (1971), an autistic math marvel in *Rain Man* (1988), and the man who figures out the secret of the Sphere in *Sphere* (1998).

Bob Hope and **Bing Crosby** have a couple of very funny math routines in *The Road to Hong Kong* (1962).

Anthony Hopkins plays a mad, brilliant mathematician working on the Riemann hypothesis in *Proof* (2005).

Elizabeth Hurley is the devil in *Bedazzled* (2000), dismissing Fermat's last theorem as useless homework.

John Hurt is a brilliant and arrogant mathematician playing detective in *The Oxford Murders* (2008).

Samuel Jackson plays with jugs in *Die Hard: With a Vengeance* (1995), and plays a brilliant mathematician in *Sphere* (1998).

Derek Jacobi plays Alan Turing in *Breaking the Code* (1996).

Danny Kaye sings and dances Pythagoras's theorem in *Merry Andrew* (1958).

Ben Kingsley plays a math professor in *Teen Patti* (2010), a Bollywood version of *21*.

Val Kilmer applies a nice reflection trick to save the day in *Red Planet* (2000). He remarks that: "This is it, the moment they told us about in high school when one day algebra would save our lives."

Lindsay Lohan is really good at math in *Mean Girls* (2004) and even joins a school math club. In *Freaky Friday* (2003) **Jamie Lee Curtis** is caught in her daughter (Lindsay Lohan's) body, and at one point is trying to remember π.

Madonna sings a very mathematical song in *Dick Tracy* (1990): "Count your blessings, one, two, three. I just hate keeping score. Any number is fine with me. As long as it's more. As long as it's more! I'm no mathematician, all I know is addition. I find counting a bore. Keep the number mounting, your accountant does the counting ..."

Tobey Maguire as Spiderman knows that Bernoulli found the curves of quickest descent in *Spider-Man 2* (2004). He also knows his eigenvalues.

Harpo Marx plays Isaac Newton in *The Story of Mankind* (1957).

The Marx Brothers can be admired in a little mathematical comedy action in *Duck Soup* (1933), *The Story of Mankind* (1957), and *The Cocoanuts* (1929).

Ewan McGregor is the main character in *Solid Geometry* (2002), who uses "a plane without a surface" to make his unwanted girlfriend disappear.

Sam Neill is a scientist in *Event Horizon* (1997), who explains, using a piece of paper, how bending space enables the spaceship *Event Horizon* to move faster than light.

Paul Newman is the rocket scientist and spy in *Torn Curtain* (1966), who uses π as a secret password and has a very mathematical discussion with an East German colleague. He also sells circles in *The Hudsucker Proxy* (1994).

Leonard Nimoy has a few very mathematical moments as Mr. Spock in *Star Trek*. He asks the computer to calculate π to the last digit in "Wolf in the Fold" (1967), and demonstrates lightning calculation in "The Trouble With Tribbles" (1967).

Donald O'Connor is the dancing actuary in *Are You With It?* (1948).

Edward James Olmos plays the legendary math teacher Jaime Escalante in *Stand and Deliver* (1988).

Gwyneth Paltrow is a mathematician in *Proof* (2005). She is also spotted with Isaac Newton's book *Mathematical Principles of Modern Philosophy* in *Sky Captain and the World of Tomorrow* (2004).

Sarah Jessica Parker has some mathematical moments in *Sex and the City*. In the episode "Ex and the City" (1999) she ponders her ex in terms of the standard mathematical unknown X. Other episodes to check out are "The Fuck Buddy" (1999), and "Old Dogs, New Dicks" (1999).

Sidney Poitier is the teacher in *To Sir, with Love* (1967), with a couple of mathematical moments.

Vincent Price plays the poet Omar Khayyam in *Son of Sinbad* (1955). Of course, Omar Khayyam was also a famous mathematician, but that is not mentioned in this movie.

Pierre Richard plays a very funny math teacher in the French comedy *Lucky Pierre* (1974).

Don Rickles in *Kelly's Heroes* (1970) poses a problem for someone else to do in their head (unfortunately): 125 boxes at \$ 8,400 a box = ... \$ 10,500,000.

Tim Robbins pretends to be a mathematician/physicist in *I.Q.* (1994) and is the main circle-loving character in *The Hudsucker Proxy* (1994).

Meg Ryan is a mathematician/physicist and Einstein's niece in *I.Q.* (1994). She knows about Zeno's paradoxes.

William Shatner has a number of very mathematical moments in *Star Trek*. He boosts a computer by a factor of "one to the power of four" in "Court Martial" (1967), and he suggests exorcising an alien from a computer by posing an unsolvable mathematical problem in "Wolf in the Fold" (1967). And in "The Trouble With Tribbles" (1967), he has trouble with tribbles, which reproduce at an exponential rate.

Kevin Spacey is a math professor in the movie *21* (2008) who organizes some of his students into a team of blackjack players.

James Spader is a cryptanalyst in *Alien Hunter* (2003) and the hero in *Supernova* (2000), a movie featuring a higher-dimensional bomb. Finally, he also comes up with a very unconvincing mathematical solution in *Stargate* (1994).

Jimmy Stewart is a pollster in *Magic Town* (1947) who discovers the ideal town for polling: its population thinks exactly like America as a whole. He is also the boffin in *No Highway in the Sky* (1951) who works on the Goldbach conjecture in his spare time.

Patrick Stewart as Captain Picard unwinds by working on Fermat's last theorem in the *Star Trek: The Next Generation* episode "The Royale" (1989). In the episode "Kryptos" (2006) of the TV miniseries *Eleventh Hour*, Stewart recognizes the Fibonacci numbers in a piece of research on climate change.

Sharon Stone is a murderous math teacher in *Diabolique* (1996).

Barbra Streisand in *The Mirror Has Two Faces* (1996) knows about prime numbers and how to handle clueless mathematicians.

Tilda Swinton is Ada Byron, Countess of Lovelace, in *Conceiving Ada* (1997).

Audrey Tautou is a cipher expert in *The Da Vinci Code* (2006), who recognizes that the numbers 13-3-2-21-1-1-8-5 are all Fibonacci numbers.

Uma Thurman runs down an Escher-like staircase and into a Pacman maze of rooms in *The Avengers* (1998).

Michelle Trachtenberg uses geometry to help her perfect her ice-skating routines in *Ice Princess* (2005). She also plays a role in *Beautiful Ohio* (2006), an excellent, dark movie about a math prodigy and his family.

Christopher Walken is a very scary angel in *Prophecy* (1995), giving some good advice: "See ya, kids. Study your math. Key to the Universe."

Denzel Washington explains the significance of X in *Malcolm X* (1992). A scientist explains to him, as the hero in *Déjà Vu* (2006), how a shortcut in space-time can be created by folding a piece of paper.

Rachel Weisz plays the mathematician Hypatia in *Agora* (2009).

Cornel Wilde plays the famous poet and mathematician Omar Khayyam in *Omar Khayyam* (1957).

Robin Williams in *Dead Poets Society* (1989) is a poetry teacher, who doesn't think much of a mathematical way of calculating the greatness of a poem. In *Flubber* (1997), he is Professor Phillip Brainard, who stumbles into a life drawing class and begins lecturing about gravity. In *Good Will Hunting* (1997), Williams is the psychologist who helps the mathematical wunderkind. And in *Patch Adams* (1998), he plays the main character, who meets a brilliant scientist/mathematician.

Bruce Willis solves a high stakes jugs puzzle in *Die Hard: With a Vengeance* (1995).

Kate Winslet as Rose in *Titanic* (1997) does some math to figure out that there are not enough lifeboats.

Reese Witherspoon is puzzled about the strange geography of *Pleasantville* (1998).

Elijah Wood is a bright young mathematician playing detective in *The Oxford Murders* (2008), and has some problems with geometry homework in *The Ice Storm* (1997).

20.7 Math Consultants

Here is a short list of mathematicians who acted as mathematics consultants for cinema movies.

Antonia's Line (1995): Wim Oudshoorn, who has a PhD in mathematics.

A Beautiful Mind (2001): Dave Bayer, professor of mathematics at Columbia University. He also acted as the hand double of Russell Crowe and makes an appearance in the movie.

Contact (1997): Tom Kuiper of the Jet Propulsion Laboratory at the California Institute of Technology and Linda Wald, then a graduate student at UCLA, were mainly responsible for what is happening on the computer screens throughout the movie.

Cube (1997): David W. Pravica, a professor of mathematics at East Carolina University.

Donald in Mathmagic Land (1959): Heinz Haber, a well known rocket scientist and science popularizer.

Good Will Hunting (1997): Patrick O'Donnell, a professor of physics at the University of Toronto. Daniel J. Kleitman, a professor of mathematics at MIT, also did some consulting in the planning stages of the movie.

It's My Turn (1980): Benedict H. Gross, a professor of mathematics at Harvard University.

The Mirror Has Two Faces (1996): Henry C. Pinkham, a professor of mathematics at Columbia University.

Sneakers (1992): Leonard Adleman, the "A" in the famous RSA encryption scheme and a professor of computer science at the University of Southern California. Check out his webpage for a nice account of his contribution to this movie.

Chapter 21
Topics Lists

For this chapter, we have compiled a number of annotated lists relating to mathematical topics:

1. Counting from 0 to 101
2. Math titles but no math
3. Pythagoras's theorem and Fermat's last theorem
4. Geometry
5. Higher dimensions
6. Topology
7. Golden ratio and Fibonacci numbers
8. Pi
9. Prime numbers and number theory
10. Chaos, fractals, and dynamical systems
11. Communicating with aliens
12. Code breaking
13. Calculus
14. Infinity
15. Paradoxes
16. Probability, gambling, and percentages
17. Famous formulas, identities, and magic squares
18. Mathematical games

21.1 Counting to 101

Zéro de Conduit (1933), One Flew Over the Cuckoo's Nest (1975), Two for the Road (1967), ¡Three Amigos! (1986), Four Weddings and a Funeral (1994), Five Easy Pieces (1970), Six Degrees of Separation (1993), Se7en (1995), Eight Below (2006), 8 1/2 (1963), The Nine Tailors (1974), 10 (1979), Ocean's Eleven (1960), 12 Angry Men (1957), Friday the 13th (1980), 7 Plus Seven (1970), 15 Minutes (2001), Sixteen Candles (1984), Number Seventeen (1932), 18 Again! (1988), Nineteen and Phyllis (1920), Twenty Bucks (1993), 21

(2008), Catch 22 (1970), The Number 23 (2007), 24 (2001), 25 Degrés en Hiver (2004), Tag 26 (2003), 27 Dresses (2008), 28 Days Later ... (2002), 29 Reasons to Run (2006), 30 Virgins and Pythagoras (1977), Kilómetro 31 (2006), Un 32 Août sur Terre (1998), Tridtsat Tri (Nenauchnaya Fantastika) (1965), Naked Gun 33 1/3 (1994), Miracle on 34th Street (1947), Kalibre 35 (2000), 36 Quai des Orfèvres (2004), To Gillian on Her 37th Birthday (1996), Quelli Della Calibro 38 (1976), The 39 Steps (1935), The 40 Year Old Virgin (2005), 41 (2007), 42nd Street (1933), Shell 43 (1916), Moon 44 (1990), Love and a .45 (1994), Code 46 (2003), 47 Morto Che Parla (1950), 48 Hrs. (1982), 49th Parallel (1941), Attack of the 50 Foot Woman (1958), The 51st State (2001), 52 Pick-Up (1986), 53 Días de Invierno (2006), Car 54, Where Are You? (1994), 55 Days at Peking (1963), Nasser 56 (1996), Passenger 57 (1992), Grenzstation 58 (1951), Psyche 59 (1964), Gone in Sixty Seconds (2000), Highway 61 (1991), Cover-Up '62 (2004), Rule Sixty-Three (1915), 64 Squares (2007), Estambul 65 (1965), Route 66 (1960), Don Juan 67 (1967), '68 (1988), The Fighting 69th (1940), Seventy (2003), 71 Fragmente einer Chronologie des Zufalls (1994), 72 Gradusa Nizhe Nulya (1976), Winchester '73 (1950), Undine 74 (1974), 75 Centilitres de Prière (1995), Segment 76 (2003), 77 Sunset Strip (1958), Seventy-8 (2004), 79 Af Stödinni (1962), Around the World in 80 Days (1956), Dancer, Texas Pop. 81 (1998), Metropolitan Police Branch 82 (1998), Gypsy 83 (2001), Nineteen Eighty-Four (1984), Airport 85 (1983), Love 86 (1986), Subject 87 (2007), Tanner '88 (1988), Junket 89 (1970), Allemagne 90 Neuf Zéro (1991), Kidô Senshi Gundam F91 (1991), 92 in the Shade (1975), United 93 (2006), 94 Arcana Drive (1994), 95 Oktaania (1990), Number 96 (1972), 97 Ga Yau Hei Si (1997), Power 98 (1996), Convict 99 (1938), One Hundred Men and a Girl (1937), One Hundred and One Dalmatians (1961).

21.2 Math Titles but No Math

Proof (1991), Murder by Numbers (2002), Zorn's Lemma (1970), The Butterfly Effect (2004), Teorema (1968), The Three Body Problem (2004).

21.3 Pythagoras's Theorem and Fermat's Last Theorem

Also check out chapter 14 ("Pythagoras and Fermat Go to the Movies").

30 Virgins and Pythagoras (1977): Czech heartthrob Karel Gott plays a math teacher who becomes a rock star performing the song "Thanks, Mr. Pythagoras."

Bedazzled (2000): The mathematical highlight is Elizabeth Hurley as a devil-teacher, dismissing Fermat's last theorem as useless.

Fermat's Last Tango (2001): Terrific musical about Fermat's last theorem, which also features Pythagoras's theorem as well as the mathematical superheroes Pythagoras, Euclid, Fermat, Gauss, Newton, and Wiles.

Furuhata Ninzaburō: "Murder of a Mathematician" (1995): Two mathematicians have just received the prestigious Australian Arbuckle award (for mathematicians under 40), for their work on dynamical systems in four-dimensional manifolds. When one is killed, the Columbo-like Furuhata solves the mystery: the other mathematician has killed him to take credit for the solution of Fermat's last theorem.

Help: "1.5" (2005): Hilarious scene of a mathematician trying to explain Fermat's last theorem to his clueless psychiatrist.

Love, Math and Sex (1997): The heroine of the movie is constantly doing math. In terms of geometry she talks about rhombuses and Pythagoras's theorem and compares her lover to a trapezium and a heptagon topped with a circle. (She's quite a catch.)

The Man Without a Face (1993): A very scary Mel Gibson quizzes a boy on Pythagoras's theorem.

Merry Andrew (1958): Danny Kaye dances Pythagoras's theorem.

The Oxford Murders (2008): In this movie we see a lecture of a professor giving the solution to "Bormat's last theorem."

The Simpsons love Pythagoras's theorem and Fermat's last theorem. In the episode "$pringfield (or, How I Learned to Stop Worrying and Love Legalized Gambling)" (1993), Homer finds some Henry Kissinger glasses, and promptly recites the Scarecrow's lines from the *Wizard of Oz*. In the episode "Homer^3" (1995) we see the following "counterexample" to Fermat's last theorem floating in midair: $1782^{12} + 1841^{12} = 1922^{12}$; see figure 21.1. In the "Wizard of Evergreen Terrace" (1998) Homer finds another "counterexample": $3987^{12} + 4365^{12} = 4472^{12}$.

Star Trek: The Next Generation: "Datalore" (1988): Commander Riker uses Pythagoras's theorem to trick the evil android Lore into revealing that he knows a lot more than he admits.

Star Trek: The Next Generation: "The Royale" (1989): Captain Picard is trying to prove Fermat's last theorem.

Star Trek: Deep Space 9: "Facets" (1995): Daks says she is working on a proof of Fermat's last theorem that is different from Andrew Wiles's.

The Wizard of Oz (1939): The Scarecrow proves to himself (but nobody else) that he's got a brain by trying to recite Pythagoras's theorem: "The sum of the square roots of any two sides of an isosceles triangle is equal to the square root of the remaining side."

Fig. 21.1 A "counterexample" to Fermat's last theorem featured in *The Simpsons*.

21.4 Geometry

Check out also the separate lists for occurrences of Pythagoras's theorem (21.3), higher-dimensional geometry (21.5), topology (21.6), the golden ratio (21.7), and π (21.8). See also chapter 5 (*"Nitpicking in Mathmagic Land"*) and chapter 14 (*"Pythagoras and Fermat Go to the Movies"*).

2010 (1984): A monolith left by aliens has proportions 1:4:9.

Amy & Isabelle (2001): A teacher talks of the beauty and usefulness of math. There is an extended classroom scene on right triangles and an extended scene on Millay's famous poem: "Euclid alone has looked on beauty bare . . ."

Au Revoir Les Enfants (1987): A teacher discusses the fact that, in a quadrilateral circumscribed around a circle, the sum of two opposite sides is equal to the sum of the two remaining sides.

The Aviator (2004): Howard Hughes passes off a meteorologist as a mathematician to a censorship board objecting to Hughes's movie. The "mathematician" uses calipers to prove that Jane Russell's décolleté is consistent with that of other movie stars.

Batman (1966): "Penguin, Joker, Riddler, and Cat Woman, too. The sum of the angles of that rectangle is too monstrous to contemplate!"

Battlefield Earth (2000): The hero is put into a learning machine and learns about simple mathematics such as equilateral triangles, the quadratic formula, and so on.

Better Off Dead (1985): A comedy in which a math teacher doing nonsensical geometry is worshipped by his students.

Cabiria (1914): Archimedes uses a compass to design his ship-burning mirrors.

Caddyshack (1980): A comedy golf scene: to illustrate that the shortest distance between two points is a straight line, Chevy Chase makes his ball jump over another one into the hole.

Cartesius (1974): More philosophy than math in Roberto Rossellini's biopic. One extended scene has Descartes solving a gravitational problem posed by the mathematician Isaac Beeckman.

Cast Away (2000): Tom Hanks calculates the area of the circular region that his rescuers will have to search.

Conceiving Ada (1997): A turgidly told story of the mathematicians Ada Byron and Charles Babbage. Several scenes have references to basic geometry: the unique parallel line, Ada's father calls her "his princess of parallelograms," the separate lives of her and her lover are compared to "parallel lines running to infinity."

Death of a Cyclist (1955): A main character in the movie is a mathematics professor. In one extended scene we see a student working on a spectacular blackboard, full of equations and drawings of rolling curves (cycloids and the like).

Diabolique (1955): This movie contains an extended scene dedicated to a boy finding the area of a regular hexagon in terms of the radius of the circumcircle.

Donald in Mathmagic Land (1959): This movie is to mathematics what Disney's *Fantasia* is to classical music. It features tons of geometry, such as the golden ratio and conic sections. See chapter 5.

The Dot and the Line (1965): An engaging animated story of the love between a dot and a line.

Dreams in the Witchhouse (2005) has references to non-Euclidean geometry. A weirdly warped corner of a room is the gateway to the witch's lair.

The Englishman Who Went Up a Hill and Came Down a Mountain (1995): Villagers are trying to raise their hill by 20 feet so that it will officially be a mountain. Once they have piled up earth to raise the height 14 feet, the pastor declares that they've broken the back of the problem. He is corrected by a villager, who apparently realizes that as long as it keeps its shape, the volume of a cone does not depend linearly on its height.

The Hudsucker Proxy (1994): The circle is used throughout as a symbol of elegance and simplicity.

The Ice Storm (1997): One boy helps another with his homework: "It's like, you know when they say "two squared"? You think it means 2 times 2, equals 4? But really they really mean a square. It's really space, it's not numbers, it's space. And it's perfect space, but only in your head, because you can't draw a perfect square in the material world, but in your mind, you can have perfect space. You know?"

The Infinite Worlds of H. G. Wells (2001): The Pyecraft episode in this miniseries is a fun fantasy about a gentle mathematician. Among many other things, he attempts to explain to a class of uninterested school children about the radius, diameter, and circumference of a circle.

The Keeper: The Legend of Omar Khayyam (2005): A movie about the life of the mathematician Omar Khayyam and one of his descendants. More romance than math, but there are some good scenes involving references to Euclid.

Lambada (1990): *Stand and Deliver* with dirty dancing. Blade is a supercool math teacher who inspires his students. Most memorable cheesy scene: Blade demonstrates the usefulness of math by making a near impossible three-cushion pool shot using "the rectangular coordinate system" and a protractor. He also teaches some basic facts about angles, right triangles, and $\sin^2 \theta + \cos^2 \theta = 1$.

The Last Casino (2004): Barnes is a math professor testing the waitress Elise for inclusion in a blackjack team. He gives her an incredibly complicated order which involves multiple toppings, substitutions, and rotation of pizza 90 degrees during construction. She gets it right.

Like Mike (2002): Calvin is being helped with geometry by the basketball player Tracey Reynolds. They paint huge triangles on the side of Tracey's house, with famous basketball players as the vertices.

Little Man Tate (1991): A movie featuring two mathematical prodigies. Lots of math: math competitions, building a tensegrity icosahedron, and so on.

Love and Death (1975): A Woody Allen movie with a couple of funny math bits: for instance, designing blintzes with geometry on a blackboard (figure 21.2).

Madame Curie (1943): Marie Curie asks her future husband a mathematical question about geometrical symmetries.

The Man Without a Face (1993): Mel Gibson as a scary math teacher quizzes a boy on Pythagoras's theorem, and teaches him a Euclidean theorem on how to find the center of a circle: he goofs it!

Fig. 21.2 Advanced blintz design in *Love and Death*.

Merry Andrew (1958): Danny Kaye plays a math teacher who teaches Pythagoras's theorem and the fact that "parallel lines never connect" by singing and dancing. See chapter 14.

Omar Khayyam (1957): Cornel Wilde is an improbable Omar Khayyam in this Hollywood biopic. A lot of poetry and romance, and a little bit of geometry.

Phase IV (1974): A mathematician communicates with intelligent ants. He sends a square and the ants respond with a circle.

Picnic at Hanging Rock (1975): One of the people who goes missing at Hanging Rock is a math teacher. We see her reading a geometry text.

The Prince and the Pauper (1990): Mickey Mouse as the prince has to sit through a very boring trigonometry lesson.

Prospero's Books (1991): One of Prospero's books is a math book featuring animated geometric drawings.

A Pure Formality (1994): A famous writer has written a book about a mathematician (Claude?) Shannon, and attributes his passion for math to one of his math teachers. He ponders projective geometry.

The Return of Sherlock Holmes: "The Musgrave Ritual" (1986): As part of a treasure hunt, Sherlock Holmes has to figure out the length of the shadow of an elm tree that has been chopped down. He knows the height of the tree and then calculates the length of the shadow from the height of a fishing rod and the length of the rod's shadow.

Rushmore (1998): The movie begins with a dream sequence in which a mathematical underachiever fantasizes that he becomes the class hero by solving a superdifficult geometry problem. The problem shown is to calculate the area of an ellipse, which is actually pretty easy.

The Siege of Syracuse (1960): Archimedes as the hero in a sword and sandals flick?! The usual geometrical constructions, and the use of parabolic mirrors to burn Roman warships and women's dresses.

Solid Geometry (2002): A man reads his grandfather's work on "a plane without a surface" and applies it to make his unwanted girlfriend disappear.

Sphere (1998): This movie features a perfect sphere as the main focus of attention. One of the main characters is a mathematician, and there's some nonsensical code breaking.

Stand-In (1937): Mathematician Atterbury Dodd applies mathematics to everything, including dancing and knocking out a bully using the principle that "the straight line is the shortest distance between two points."

Succubus (1968): A bizarre scene in which someone is playing the piano from a geometry book while a woman is taking off her clothes.

Star Trek: The Next Generation: "The Vengeance Factor" (1989): Wesley Crusher explains his homework: "This is the locally Euclidean metrization of a k-fold contravariant Riemannian tensor field."

Waterboys (2001): Some schoolboys take up synchronized swimming. One is a math wiz who uses geometric diagrams to plan one of their routines.

21.5 Higher Dimensions

Also check out chapter 15 ("Survival in the Fourth Dimension"), which is dedicated to this topic.

Back to the Future Part III (1990): Doc realizes he has to "think four-dimensionally" in order to drive a locomotive toward a bridge that hasn't been built yet.

Cube 2: Hypercube (2002): A bunch of victims in a deadly maze, supposedly a hollow hypercube. See chapter 15.

Déjà Vu (2006): There's a scene in which a shortcut in space-time is demonstrated by folding a piece of paper.

Event Horizon (1997): The movie contains a nice scene in which a scientist folds a piece of paper to explain how bending space enables their spaceship to move faster than light.

Flatland (1965): A nice animated adaptation of Edwin Abbott's famous book, *Flatland*, about a two-dimensional world.

Flatlandia (1982): Michele Emmer's claymation adaptation of Abbott's famous book.

Flatland: The Film (2007): Yet another adaptation of Abbott's book. This one is pretty tedious.

Flatland: The Movie (2007): One more, state of the art, animated adaptation of Abbott's *Flatland*. Toward the end of the movie we see a rotating hypercube slicing through a three-dimensional world.

Insignificance (1985): Albert Einstein and Marilyn Monroe have an exchange of ideas. Marilyn demonstrates the theory of relativity using toy trains and flashlights. Einstein explains to Marilyn (not quite correctly) the shape of the Universe as a three-dimensional sphere.

Outer Limits: "Behold Eck!" (1964): In this great episode, a two-dimensional alien gets stranded on Earth, walks through walls (sideways!), and slices through buildings (figure 21.3).

Fig. 21.3 The two-dimensional alien in the *Outer Limits* episode "Behold Eck!"

Shrieker (1998): A very bad horror movie, featuring a higher-dimensional monster and a math student specializing in "multidimensional topography."

The Simpsons: "Homer^3" (1995): Homer stumbles into a 3D world, filled to the brim with mathematical bits and pieces.

Supernova (2000): The crew of a spaceship come across a bomb made from higher-dimensional matter. Very nice graphics (correct up to a certain point), which show how to build two-dimensional projections of higher-dimensional cubes.

Threshold: "Trees Made of Glass" (2005): A four-dimensional alien object intersects the world, driving people insane.

21.6 Topology

Lots of Möbius strips, Pacman spaces, and Escher-like constructs in this list. Also have a look at list 21.5 on higher dimensions and chapter 15 ("Survival in the Fourth Dimension").

Antonia's Line (1995): A movie about a wunderkind who becomes a mathematician. Wunderkind calculations, and a homological algebra lecture with commutative diagrams on blackboards.

The Avengers (1998): Uma Thurman runs down an Escher-like staircase and into a Pacman maze of rooms.

Graveyard Disturbance (1987): Some teenagers are trying to escape from a cemetery that loops in on itself in an impossible way. They discuss it rather weirdly, in terms of Escher and non-Euclidean geometry.

Infinity (1996): A movie about Richard Feynman featuring the Möbius strip, an abacus, and an explanation of infinity. He uses the Möbius strip to impress his girlfriend.

Insignificance (1985): Albert Einstein and Marilyn Monroe have an exchange of ideas. Albert explains (not quite correctly) the shape of the Universe as a three-dimensional sphere.

It's My Turn (1980): Light romantic movie about a female mathematician. Famous among mathematicians because the full proof of the snake lemma is given in the first scene. There are other bits and pieces of homological algebra on blackboards. See chapter 12.

Izo (2004): A movie featuring a guy with a samurai sword running around on a Möbius strip.

Love, Math and Sex (1997): The heroine is a mathematical prodigy. The nonsex parts include quadratic equations, probability, topology, the Möbius strip, and a math competition.

The Matrix Revolutions (2003): Neo gets stuck in a train station which loops around à la Pacman.

Moebius (1996): A subway train goes missing in the subway network of Buenos Aires. A topologist is called in to investigate. He concludes that a new section of the subway network is acting like a Möbius strip (whatever that means).

Pinky and the Brain: "The Maze" (1997): Pinky and the Brain are put into a hologram maze, which has a terrific Escher section in it.

Pleasantville (1998): In the sitcom world of Pleasantville, a geography teacher explains how the town loops back on itself. Very funny scene, but on the DVD commentary the writer-director shows he's completely confused.

The Spanish Prisoner (1997): Intrigue centered around a valuable industrial process. Bits and pieces of math as decoration: a painting of Luca Pacioli, a commutative diagram, multivariable calculus, and so on.

21.7 Golden Ratio and Fibonacci Numbers

Check out also chapter 4 ("The Annotated Pi Files"), chapter 5 ("Nitpicking in Mathmagic Land"), and chapter 16 ("To Infinity and Beyond").

21 (2008): A math professor organizes some of his students into a team of blackjack players. The Fibonacci numbers make a brief appearance.

After Midnight (2004): The lead characters win the lottery by playing the first few Fibonacci numbers. They spout the usual lines about Fibonacci numbers in nature.

Breaking the Code (1996): Alan Turing talks about the appearance of Fibonacci numbers in pinecones.

Brutal (2007): A mild-mannered science teacher uses flowers and Fibonacci numbers to plan his gruesome murders.

The Da Vinci Code (2006): The Fibonacci numbers appear in various scenes.

Donald in Mathmagic Land (1959): In this Disney classic, the golden ratio in nature and geometry features prominently, sometimes correctly. See chapter 5.

Eleventh Hour (2006): In the episode "Kryptos," a brilliant meteorologist uses the Fibonacci numbers to model climate change.

Mr. Magorium's Wonder Emporium (2007): Dustin Hoffman as Mr. Magorium quizzes his potential accountant (whom he dubs "Mutant"): "Name the Fibonacci series from its 11th to its 16th integer." "Uh, 89, 144, 233, 377, 610?" "Perfect! The number 4, do we really need it?" "If you like squares, you do." "Oh, I like squares!" Of course, Mutant actually only went up to the 15th Fibonacci "integer."

π (1998): The mathematical prodigy Max is looking for patterns in nature and in the stock market. Both the Fibonacci numbers and the golden ratio feature prominently. See chapter 4.

Taken: "God's Equation" (2002): It turns out God's equation is the Fibonacci sequence "writ large across the Heavens." This coming from two guys who just exploded a hamster. These same guys are trying to use Fibonacci numbers to track aliens: apparently there are 55 breeding pairs, 46,368 aliens in total, and so on. Fibonacci numbers are briefly mentioned again in the later episode "John," where people are trying to figure out the alien language.

21.8 Pi

Also check out chapter 4 ("The Annotated Pi Files").

Doctor Who: "The Five Doctors" (1983): The Doctor figures out that the key to crossing a deadly chessboard lies in the digits of π.

Doctor Who: "Midnight" (2008): Sky is taken over by some intelligence, immediately repeating whatever anyone else says. The Doctor tests her accuracy, with Sky repeating a half a second behind him: "The square root of pi is 1.77245385090551602729816748334l wow."

Donald in Mathmagic Land (1959): A little birdlike creature recites some digits of the decimal expansion of π (and messes it up). See chapter 5.

The Last Casino (2004): George, a candidate for a blackjack team, spends his leisure time memorizing the digits of π.

Never Been Kissed (1999): Very catchy mathy song is playing while we see the Denominators, members of a math club in math action. The movie includes a hilariously bad expansion of π; see figure 18.2.

Northern Exposure: "Nothing's Perfect" (1992): This episode features a mathematician who is crazy about her pets and is writing a dissertation on π. She knows that π is transcendental, and she's looking for patterns in its decimal expansion.

π (1998): The mathematical prodigy Max is looking for patterns in nature, π, and the stock market. See chapter 4.

Red Planet Mars (1952): While biting into a pie a boy has the idea of using the decimal expansion of π to communicate with aliens.

The Simpsons love π. In the episode "Bye, Bye Nerdie" (2001), Professor Frink is trying to gain the attention of the audience in a big lecture theatre: "Scientists—Scientists, please! I'm looking for some order. Some order, please, with the eyes forward and the hands neatly folded and the paying of attention—Pi is exactly three!" In the episode "Marge in Chains" (1993), Apu says: "In fact I can recite pi to 40,000 places. The last digit is one!" Homer: "Mmm, pi(e)." In "Lisa's Sax" (1997), two girls at a gifted school are reciting the digits of π as a skipping rhyme: "Cross my heart and hope to die, here's the digits that make pi: 3.1415926535897932384 ..." Finally, in "Simple Simpson" (2004), Homer, disguised as Pie Man, throws a pie at Rich Texan. Someone comments: "We all know 'pi r squared,' but today, 'pie are justice.' I welcome it."

Star Trek: "Wolf in the Fold" (1967): Spock cures a computer possessed by an alien by commanding it to calculate π to the last digit.

Torn Curtain (1966): A Hitchcock movie in which Paul Newman plays an American scientist supposedly defecting to the East. Great scene where the mathematical theories are competing. Lots of calculations on blackboards, and the symbol π is used as a secret sign.

Twilight (2008): Bella is trying to figure out Edward the vampire: "You gotta give me some answers." "Yes—No—To get to the other side—1.77245—" "I don't want to know what the square root of pi is." "You knew that?"

The Virgin Suicides (1999): Teacher in action with Venn diagrams. One of the blackboards is adorned with a messed-up decimal expansion of π. In another scene we see a blackboard with problems relating to finding the derivative of a quadratic function.

21.9 Prime Numbers and Number Theory

See also list 21.3 on Pythagoras's theorem and Fermat's last theorem. The chapters that have more details on this topic are chapter 2 ("The Beautiful Hand Behind a Beautiful Mind"), chapter 6 ("Escape from the Cube"), chapter 11 ("One Mirror has Two Faces, Two Mirrors have ..."), and chapter 18 ("Money-Back Bloopers").

A Beautiful Mind (2001): A movie about the brilliant mathematician John Nash. For the second half of the movie he is working on the Riemann hypothesis. See chapter 2.

Chen Jingrun (2001): A nine-hour soapie about the famous Chinese mathematician, documenting Chen's simultaneous struggles with authoritarian thugs and the Goldbach conjecture.

Contact (1997): SETI researchers detect an alien signal. The various layers of information are accessible via prime numbers and a cube.

The Core (2003): The first few prime numbers (starting with 1!) are used to encrypt and decrypt a secret message.

Cube (1997): Six people wake up in a deadly maze, full of mathematical clues. Factoring numbers and recognizing when they're prime turn out to be of vital importance for the group's survival. See chapter 6.

Doctor Who: "42" (2007): To get through the security door during an emergency, the Doctor's sidekicks have to type the next number in the sequence: 313, 331, 367. The Doctor immediately realizes the answer is 379: "It's a sequence of happy primes—Any number that reduces to 1 when you take the sum of the square of its digits and continue iterating until it yields 1 is a happy number—A happy prime is a number that is both happy and prime. Now type it in! I dunno, talk about dumbing down. Don't they teach recreational mathematics anymore?"

Fermat's Room (2007): Four mathematicians are trapped in a shrinking room. The Goldbach conjecture plays a major role. See chapter 7.

Flash Gordon (1980): Dr. Zarkov saves the day by guessing the code of the elevator to be "one of the prime numbers of the Zeeman series."

Futurama: "The Lesser of Two Evils" (2000): The robots Bender and Flexo share a laugh over their serial numbers, that both are expressible as the sum of two cubes: $3,370,318 = 119^3 + 119^3$ and $2,716,057 = 952^3 + (-951)^3$.

High School Musical (2006): The (high school!) teacher is writing up two of Ramanujan's formulas for $\frac{1}{\pi}$, Pochhammer symbols and all. Vanessa immediately spots an error in the second equation, where the teacher has written $\frac{8}{\pi}$ instead of $\frac{16}{\pi}$.

The Infinite Worlds of H. G. Wells (2001): The Pyecraft episode in this mini-series is a fun fantasy about a gentle mathematician. At some point it is mentioned that he solved the Riemann hypothesis.

Inspector Lewis (2006): The Goldbach conjecture and a Fields medalist are at the center of a murder mystery.

It's My Turn (1980): A light romantic movie about a female mathematician. See chapter 12. Toward the end of the movie the mathematician talks to a kid about prime numbers.

The Mirror Has Two Faces (1996): A romantic comedy about a math professor who is dragged out of his shell. Lots of references to prime numbers and the twin prime conjecture. See chapter 11.

Mozart and the Whale (2005): A charming romantic movie about Donald and Isabelle, two people with Asperger's syndrome. Donald has very strong mathematical abilities (figure 21.4). There are some touching and funny scenes about mathematical literalism and factoring license plate numbers.

The Music of Chance (1993): Two guys win the lottery by choosing primes.

Proof (2005): Gwyneth Paltrow is a troubled mathematician, coming to terms with the death of her troubled mathematician father. Both work on the Riemann hypothesis.

Sex and the City: "The Fuck Buddy" (1999): "You dumped him. Fits a pattern." "I don't have a pattern." "In math, randomness is considered a pattern." "Yes, and I'm what they call a prime number."

Sneakers (1992): A mathematician's codebreaking device becomes the inspiration for spy games. At some point he gives a lecture: "While the number-field sieve is the best method currently known, there exists an intriguing possibility for a far more elegant approach. Here we would find a composition of extensions, each Abelian over the rationals, and hence contained in

Fig. 21.4 Donald firing prime numbers at 589 in *Mozart and the Whale*.

a single cyclotomic field. Using the Artin map, we might induce homomorphisms from the principal orders in each of these fields that ... These maps could then be used to combine splitting information from all the fields." See figure 21.5.

Fig. 21.5 Dramatically projected number fields in *Sneakers*.

Threshold: "Trees Made of Glass, Part 2" (2005): The mathematician refers to "isomorphic group therapy," monotonic null-sequences, and quadratic reciprocity.

21.10 Chaos, Fractals, and Dynamical Systems

Many movies refer to chaos and the like in some vague manner. Here, we list movies that provide a little more substance with the reference.

The Bank (2001): A math prodigy takes revenge on a bank. The Mandelbrot set and chaos theory feature prominently.

Chaos (2005): A criminal who calls himself Lorentz leaves clues involving chaos theory (invented, according to the movie, by Edward Lorentz).

Drawing Down the Moon (1997): A crazy Z-grade movie, with a mathematician as the bad guy and a flaky witch character as the good guy. A few fractals appear, for reasons that are impossible to discern.

The Favor (1994): A fun romantic comedy with a mathematician as the husband (and Brad Pitt as the "favor"). Very funny scene of his wife trying to seduce him as he rabbits on about nonlinear dynamics. Later, he plays the blues on harmonica while working on some fractals.

Julie Johnson (2001): Julie Johnson is a downtrodden white trash housewife who turns out to be a mathematical genius. Tedious and humorless, but lots of math, mostly chaos theory. Mischa Baton plays the daughter. She has the Cantor set explained to her, and is less than impressed.

Jurassic Park (1993): Jeff Goldblum as the cool math dude. He impresses a woman with an explanation of chaos theory, by conducting a little experiment on her hand that illustrates the butterfly effect.

Jurassic Park II (The Lost World) (1997): Jeff Goldblum is back as the cool mathematician. No math this time.

Jurassic Park III (2001): Jeff Goldblum does not appear this time, but he and his book on chaos theory are made fun of.

Magik and Rose (1999): A square-dance caller and snail-egg farmer also turns out to be a math wiz, who knows everything about fractals and chaos theory.

Threshold (2005): This short-lived TV series contains a very funny mathematician: an acid-tongued dwarf who likes to hang out in strip joints. A particular fractal makes ominous appearances throughout the series.

21.11 Communicating with Aliens

Here we list only those instances in which the aliens don't happen to speak English.

Alien Hunter (2003): The main character is an expert cryptanalyst. Very funny blooper involving the chances of something being 99 to the infinite— but not 100.

Contact (1997): SETI researchers detect an alien signal. The various layers of information are accessible via prime numbers and a cube.

Forbidden Planet (1956): Dr. Morbius has mastered the language of an extinct civilization, starting with geometrical theorems.

Kolchak: "They Have Been, They Are, They Will Be" (1974): A pretty funny story of using an artificial language for cosmic communication, to try to communicate with the wrong type of alien: "Your present location is the third planet of our star system. We are peaceful. I will now address you in Mathmatico, the universal language—AGGHH!!"

The Man from Planet X (1951): Geometry is suggested as the way to communicate with an alien.

Phase IV (1974): A mathematician communicates with intelligent ants by sending a square. They answer with a circle.

Red Planet Mars (1952): While biting into a pie, a boy has the idea of using the decimal expansion of π to communicate with aliens.

Sphere (1998): This movie features a mysterious perfect sphere from outer space as the main focus of attention. One of the main characters is a mathematician, and there's some nonsensical code breaking.

21.12 Code Breaking

See also the previous list 21.11 on communicating with aliens.

The Amateur (1981): A spy is writing a book on Elizabethan codes and ciphers.

A Beautiful Mind (2001): A movie about the brilliant mathematician, John Nash, who won the Nobel Prize in economics for his work in game theory. In the first part of the movie John Nash is shown working as a cryptanalyst. See chapter 2.

Breaking the Code (1996): A great movie about the mathematician and logician Alan Turing who was one of the main people behind cracking the Enigma cipher.

Cube (1997): Six people wake up in a deadly maze, full of mathematical clues. See chapter 6.

The Da Vinci Code (2006): The heroes of this movie recognize the Fibonacci sequence as a key to various puzzles.

The Day of the Beast (1995): The three heroes are trying to piece together a message from the devil consisting of some letters written on individual pieces of paper: "My God." "There are hundreds of combinations." "Thousands of millions. There are 15 letters. A permutation of 15 elements with three letters repeated twice, and two letters repeated three times, giving us a total of 4,540,536,000 possibilities" (which is correct). Meanwhile, the dumbest of the three characters figures out the message.

Deadly Friend (1986): A robot tries all possible three-number combinations of numbers ranging from 0 to 40, with his creator commenting on the number of such combinations.

Enigma (2001): This movie is about Bletchley Park and the Enigma machine. The main character is modeled after Alan Turing, but there's not much math.

First Circle (1992): One of the main characters is a mathematician in a special Soviet prison for scientists. A subplot consists of the people in charge of the prison trying to recruit the mathematician into a code-breaking unit headed by his former professor.

Hands of a Murderer (1990): The battle between Sherlock Holmes and Moriarty is partially framed in terms of mathematics and codes. In the end, the key to the code involves no real mathematics.

Lost Souls (2000): A possessed man keeps writing numbers, the name of the antichrist in code.

A Man Called Intrepid (1979): Two characters talk about the Enigma machine.

Sebastian (1968): A mathematician is in charge of a special code-breaking unit, consisting exclusively of beautiful young women.

Sekret Enigmy (1979): A Polish movie about the (genuinely neglected) Polish contribution to breaking the Enigma cipher in WWII. It features a calculation of the astronomical number of different settings of the Enigma machine.

Sneakers (1992): A mathematician's code-breaking device becomes the inspiration for spy games.

21.13 Calculus

A decent amount of calculus can be found in chapter 1 ("Good Math Hunting"), chapter 3 ("Escalante Stands and Delivers"), and chapter 11 ("One Mirror Has Two Faces, Two Mirrors Have …").

21 (2008): A math professor organizes some of his students into a team of blackjack players. Newton's method makes a brief appearance.

Another Gay Movie (2006): A classroom scene like no other. Andy fantasizes having sex during a math class with his teacher, Mr. Puckov. Averting our eyes, we noticed the surface integrals on the blackboard behind them.

Battle of the Worlds (1961): A crazy/brilliant mathematician does calculus on flowerpots: "You and the others have to see and hear before you can know. I have one advantage over all of you: Calculus."

Cloak and Dagger (1946): Gary Cooper is in hiding, and passes the time by drawing on a wall, calculating the line integral of a sine curve; see figure 20.6 for a reconstruction of his calculation.

Colossus: The Forben Project (1970): Two supercomputers are figuring out a way to communicate with each other starting with times tables, getting to calculus, and then going much, much further.

Dangerous Sex Date (2001): A movie consisting of 89 minutes of sex and murder, and 2 seconds of Riemann Christoffel symbols.

Death of a Neapolitan Mathematician (1992): Story of the suicide of the Italian mathematician Renato Caccioppoli. There are some nice calculus scenes, and the sandwich theorem makes an appearance; see figure 21.6.

Fig. 21.6 A disillusioned Renato Cacciopoli teaches the sandwich theorem in *Death of a Neapolitan Mathematician*.

Futurama: "Bender's Big Score" (2007): The Harlem Globetrotters use their "razzle-dazzle globetrotter calculus" (variation of parameters and expansion of the Wronskian) to prove the possibility of paradox-free time travel.

A Hill on the Dark Side of the Moon (1983): A dark movie about the mathematician Sonya Kovalevskaya, featuring some lecture theatre scenes;

see figure 21.7. The mathematicians Weierstrass and Mittag-Leffler make brief appearances.

Fig. 21.7 Sonya Kovalevskaya lecturing in *A Hill on the Dark Side of the Moon*.

An Innocent Love (1982): A young boy tutors a school senior in calculus. Well, actually, he only seems to teach her how to differentiate polynomials.

The Last Enemy (2008): Dr. Stephen Izard explains his research in the first episode of this miniseries: "Let *Ric* be the Ricci curvature, and the gammas are the Christoffel symbols. The idea is to show that under the Ricci flow, positive curvature tends to spread outwards, uh, until at infinite time, the manifold will achieve constant curvature. Now let us consider a generalization of this, the form *F* is a nice, smooth function. Even in this state the flow will develop singularities in finite time. But, by using a local version of the Gromov compactness theorem, I've been able to get a model of all possible singularities." This is actually real mathematics, underlying Grigori Perelman's proof of the Poincaré conjecture.

Mean Girls (2004): The main character is a very pretty girl who is also very good at math. Encouraged by her teacher and against the advice of her friends ("social suicide"), she joins the school math club, takes part in a math competition, and is called to find:

$$\lim_{x \to 0} \frac{\ln(1-x) - \sin(x)}{1 - \cos^2(x)}.$$

There are also a few good classroom scenes with limits and differential equations adorning the blackboards.

The Mirror Has Two Faces (1996): A romantic comedy about a math professor who is dragged out of his shell. Tons of math. This movie is definitely one of the best sources of calculus-related movie clips: implicit differentiation, the chain rule, a calculus proof that $e^{x+y} = e^x e^y$, and more. See chapter 11.

My Teacher's Wife (1995): The opening titles feature lots of integrals in the background. In one of the first scenes, the mean teacher comments on the outcome of a calculus test: "Calculus is not for wimps."

Rushmore (1998): The movie begins with a dream sequence in which a mathematical underachiever fantasizes that he becomes the hero of the class, by solving a superdifficult math problem. The problem shown is to calculate the area of an ellipse, which is a pretty easy calculus exercise; see figure 21.8.

Fig. 21.8 Working on the "hardest geometry equation in the world" in *Rushmore*.

The Seven-Per-Cent Solution (1976): Sherlock Holmes: "The only time Professor Moriarty really occupied the role of my nemesis was when it took him three weeks to make clear to me the mysteries of elementary calculus."

She Wrote the Book (1946): A woman math professor lectures about the radius of convergence of complex power series.

Stand and Deliver (1988): A great movie about the legendary math teacher Jaime Escalante, motivating his students in a poor school in East Los Angeles. At some point Escalante decides to prepare the class for the Advanced Placement mathematics exam, which consists primarily of calculus. In the movie he shows his students how to calculate the volume of surfaces of revolution, integration by parts, and so on. See chapter 3.

Strangers on a Train (1951): A Hitchcock movie, which includes a very funny scene with a drunk mathematician discussing calculus.

Torn Curtain (1966): A Hitchcock movie in which Paul Newman plays an American scientist supposedly defecting to the East. Great scene where the mathematical theories are competing with lots of differential equations on blackboards.

21.14 Infinity

See also list 21.15 on paradoxes, list 21.6 on topology, as well as the entries featuring the phrase "parallel lines never meet" in list 21.4 on geometry. Also check out chapter 16 ("To Infinity and Beyond").

Alien Hunter (2003): The main character is an expert cryptanalyst. Very funny blooper involving the chances of something being 99 to the infinite— but not 100.

The A-Team: "The Rabbit Who Ate Las Vegas" (1983): A mathematician figures out a system to win at gambling, using "the infinity concept of declining numbers."

The Bank (2001): A math prodigy takes revenge on a bank. Tons of math. The infinity sign appears in several scenes. In particular, in one scene the infinity sign appears on the screens of the computers of the bank, signalling that the losses of the bank have just gone beyond what the computers can display.

A Beautiful Mind (2001): A movie about the brilliant mathematician John Nash. A beautiful scene, where Nash rides his bicycle in the form of an infinity sign; see figure 16.7.

Bill & Ted's Bogus Journey (1991): Bill and Ted (Keanu Reeves) run into Colonel Oats, who makes them do infinity push-ups. They contemplate whether it might be possible if they are allowed to do them girly style.

Drowning by Numbers (1988): In this Peter Greenaway movie we see a girl counting from 1 to 100 while jumping rope. She is actually counting stars and stops at 100 because "once you've counted 100, all the other hundreds are the same." Signs counting from 1 to 100 appear throughout the movie.

Forbidden Planet (1956): Dr. Morbius has mastered the language of an extinct civilization. At one point he illustrates very nicely their immense power resources, showing off a huge exponentially calibrated sequence of meters.

Hotel Hilbert (1996): David Hilbert's vision of infinities is brought to life in this brilliant educational film. The movie features a version of Zeno's arrow paradox, Thompson lamps, an Archimedes-like character counting all the grains of sand in the world, and many other puzzles of infinity.

Infinity (1996): A movie about Richard Feynman, featuring a very interesting discussion of infinity.

The Mirror Has Two Faces (1996): A romantic comedy about a math professor who is dragged out of his shell. At one point, frustrated by a student, he exclaims: "Don't you know that it is possible to remove an infinite number of elements from an infinite set and still have an infinite number of elements left over?"

Moebius (1996): A subway train goes missing in the subway network of Buenos Aires. A topologist is called in to investigate. He concludes that the new section is acting like a Möbius band (whatever that means) and remarks upon the newly infinite nature of the subway network (whatever that means).

The Phantom Tollbooth (1970): When Milo visits Digitopolis, its ruling Mathemagician convinces him of the infinity of numbers by adding 1 again and again. Earlier, the Mathemagician sings a version of Zeno's paradox, suggesting that Milo keep dividing his troubles in 2 until they disappear.

π (1998): The mathematical prodigy Max is looking for patterns in nature, the stock market, and the infinite decimal expansion of π. In one scene a little girl asks him to divide seventy-three by twenty-two. His answer: "Three point three one eight one eight one eight ..." His voice trails off into the distance as he keeps walking down the stairs. He keeps saying "one" on one step and "eight" on the next,—a very nice portrayal of the infinite decimal expansion of this number. See chapter 4.

The Saragossa Manuscript (1965): Some noblemen ponder the nature of infinity.

Star Trek: "Wolf in the Fold" (1967): Spock cures a computer possessed by an alien, by commanding it to calculate π to the last digit.

21.15 Paradoxes

Also see the various paradoxical Escher-like constructs contained in list 21.6 on topology.

Liar Paradox and Puzzle

Bedazzled (1967): Peter Cook, as the devil, explains to Dudley Moore that everything he's ever said to him is a lie. A very funny display of the liar paradox.

The Enigma of Kaspar Hauser (1974): A pompous professor presents Kaspar Hauser with a misworded version of the liar puzzle: given one question, how do you tell whether a person is a truthteller or a liar? Kaspar's simple and hilarious solution is to ask the person if they are a tree-frog. The professor is not amused.

Fermat's Room (2007): Four mathematicians are trapped in a shrinking room that can only be stopped from shrinking by solving a number of puzzles. The liar puzzle is one of the problems that the mathematicians have to solve. See chapter 7.

Labyrinth (1986): A fun fantasy version of the liar puzzle: deciding which of two doors to choose when one is guarded by a truthteller and one by a liar.

Zeno's Paradoxes

I.Q. (1994): Meg Ryan and Tim Robbins provide a romantic illustration of one of Zeno's paradoxes; see page 180.

Jam Films 2: Armchair Theory (1994): In a fantasy sequence we see one Japanese warrior shoot an arrow at a second warrior who is running away, illustrating another of Zeno's paradoxes: when the arrow has reached the point at which the second warrior was when the arrow was fired he has already moved to a second point, when the arrow reaches that point he has moved to a third point, and so on (figure 21.9).

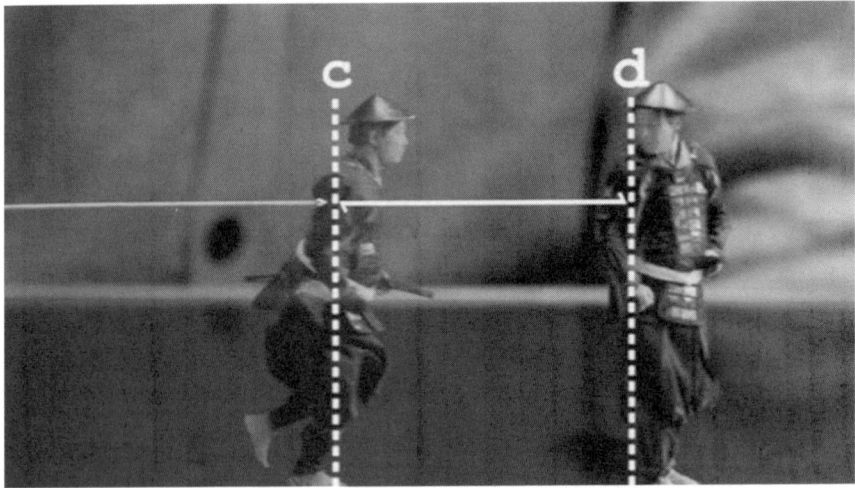

Fig. 21.9 Zeno's paradox, as a warrior fleeing an arrow in *Armchair Theory*.

The Phantom Tollbooth (1970): When Milo visits Digitopolis, its ruling Mathemagician convinces him of the infinity of numbers by adding 1 again and again. Earlier, the Mathemagician sings a version of one of Zeno's paradoxes, suggesting that Milo keep dividing his troubles in 2 until they disappear.

Self-Reference

The Infinite Worlds of H. G. Wells (2001): The Pyecraft episode in this mini-series is a fun fantasy about a gentle mathematician Pyecraft. At some point Pyecraft notes that: "There is actually nothing very interesting about the number 39. But of course when one thinks about it in this way makes it an especially interesting number, because it is the smallest number having the property." "The smallest number having the property of being uninteresting. So, it's interesting and uninteresting at the same time."

21.16 Probability, Gambling, and Percentages

12 Angry Men (1957): Probability is used by a jury member to cast doubt on the accused's guilt.

Aces (2006): Math majors, one with the alias "Theorem Girl," spend the summer playing poker. Their probability isn't quite right.

A-Ducking They Did Go (1939): The three stooges have a funny one-liner on percentages.

Alien Hunter (2003): The main character is an expert cryptanalyst. This movie includes a funny blooper involving the chances of something being 99 to the infinite—but not 100.

The Arrival (1996): There is a very funny blooper involving an incorrect way to add percentages.

The Bridge on the River Kwai (1957): "If you make one [parachute] jump you've got 50% chance of injury. If you make two, 80%, and three you are bound to catch a packet.—Go ahead and jump and hope for the best," Response from the jumper: "With or without parachute?"

Chicago (2002): "In 47 years, Cook County ain't never hung a woman yet. So, the odds are 47 to 1 that they won't hang you."

Class Action (1991): An actuarial bean-counter shows why it is cheaper to deal with lawsuits than to fix a deadly fault in some cars.

Jason X (2001): A cute android named Kay-Em calculates the probability of the team's survival at 12%. After her friend Tsunaron kisses her, she calculates the odds at 53%. He suggests they try for 100%.

Las Vegas Shakedown (1955): A math teacher travels to Las Vegas to do research for her book, which is supposed to show that the casinos always win (duh). Some humorous lines about a companion's gambling system.

Las Vegas Weekend (1986): A crazy B-grade comedy about a mathematician with a scheme to beat blackjack. Some weird and humorous lines about infinity and probability.

Loser Takes All! (2003): A math teacher claims to have found a system for predicting the lottery and he proves it by winning the jackpot twice. An expert gambler with a perfect memory is called in by the police to find out what is going on. A fun movie with lots of math, gambling, and mathematically minded people.

Magic Town (1947): Somebody locates the ideal town for polling: its population thinks exactly like America as a whole.

My Little Chickadee (1940): A very funny one-liner from W. C. Fields about whether poker is a game of chance: "Not the way I play it, no."

Rosencrantz & Guildenstern Are Dead (1990): Rosencrantz and Guildenstern have a coin come up heads 157 times in a row. As they keep tossing the coin, they discuss what this means.

The Rules of Attraction (2002): A very funny scene where a girl decides that wearing two condoms is 196% safe.

Sex and the City: "Old Dogs, New Dicks" (1999): "Hey. If 85% aren't circumcised, that means I've only slept with 15% of the population. Tops." "Wow. You're practically a virgin."

Threshold: "Outbreak" (2006): The mathematician Ramsey is "determining the corresponding probability characteristics of a system of random variables" (by asking people in a market how many tomatoes they just bought and how many people make up their family.)

Willy Wonka & the Chocolate Factory (1971): Some funny scenes involving percentages. One has a teacher using Wonka Bars to teach percentages. Another has a total of 105%, and a computer is programmed to find golden tickets based on the "computonian law of probability."

21.17 Famous Formulas, Identities, and Magic Squares

Also check out list 21.3 on Pythagoras's theorem and Fermat's last theorem for $a^2 + b^2 = c^2$ and $x^n + y^n = z^n$ and chapter 14 ("Pythagoras and Fermat Go to the Movies").

$1 + 1 = 2, 2 + 2 = 5$

$1 + 1 = 2$ is one of the identities floating around in the strange mathematical world that Homer stumbles into in *The Simpsons* episode "Homer[3]" (1995). A person is brainwashed into believing that $2 + 2 = 5$ in *1984* (1984), and in *Ball of Fire* (1941) academics argue over whether the correct grammar is "two and two is five" or "two and two are five." In *The Gambler* (1974) $2 + 2 = 5$ is used as a metaphor for the gambler's belief that he can overcome the odds. In *The Road to Hong Kong* (1962) Bing Crosby tests Bob Hope's memory by asking him what $2+2$ is. Finally, the gangster movie *Force of Evil* (1948) features an accountant called Two and Two, who is great at mental arithmetic.

Times Tables

In *Colossus: The Forben Project* (1970) two supercomputers are figuring out a way to communicate with each other—starting with the times tables. In *Box of Moon Light* (1996) a very nerdy father prescribes gigantic flash cards to his son for learning the times tables. In *Fahrenheit 451* (1966), schoolchildren drone the nine times table. Some singing of times tables can also be admired in *Calabuch* (1956).

$e^{\pi i} + 1 = 0$

Euler's identity, one of the most beautiful in mathematics, is one of the identities floating around in the strange mathematical world that Homer stumbles into in *The Simpsons* episode "Homer[3]" (1995). *The Professor and His Beloved Equation* (2006) is a beautiful movie, the story of a mathematics professor who only remembers things for a short time. He likes perfect and amicable numbers and his favorite equation is Euler's identity.

$\sin^2 \theta + \cos^2 \theta = 1$

This identity makes an appearance in *Lambada* (1990).

Quadratic Formula

The quadratic formula is one that the hero learns from a machine in *Battlefield Earth* (2000). In *Love, Math and Sex* (1997) the wizkid applies this formula on several occasions. The quadratic formula is mentioned in the famous math song in *The Pirates of Penzance* (1983) "I am the very model of a modern Major-General ... I'm very well acquainted, too, with matters mathematical. I understand equations, both the simple and quadratical. About binomial theorem I'm teeming with a lot o'news ..." In the *Family Guy* episode "Let's Go to the Hop" (2000) there is a flashback to a pilgrim correctly reciting the quadratic formula. She is praised; then, being a girl who can answer math problems, she is declared a witch.

Binomial Formula

The famous math song in *The Pirates of Penzance* (1983) mentioned above also contains the binomial formula. While the kids are making out in *School-*

girl Report 4 (1972), the mathematically minded viewer will be sidetracked by the binomial formula on the blackboard.[1]

Magic Squares

Bianca (1984) has a classroom scene with Dürer's famous magic square as the centerpiece. This special magic square also appears in the background in one of the scenes in *Münchhausen* (1943). In the kung fu flick *Brave Archer 3* (1981), the properties of magic squares are explained by the heroine using the usual 3 × 3 magic square. She also uses these properties to complete a 10 × 10 magic square (figure 21.10).

Fig. 21.10 Two magic squares in Chinese characters from *Brave Archer 3*, and one from *Bianca*. In the 10 × 10 magic square all the rows, columns, and diagonals add up to 505. In the movie, the circled 63 is mistakenly drawn as a second 62.

The Circumference and Area of a Circle

The latter allows Tom Hanks in *Cast Away* (2000) to calculate the circular area that his rescuers will have to search. In the Pyecraft episode of *The*

---

[1] eBay has been one of our methods of finding obscure math movies, with unpredictable results: we're not always sure what we're buying. In the case of *Schoolgirl Report 4*, we certainly did not predict that the video would be seized by Australian customs. Confirming with the eBay seller that there was nothing to worry about, we contested the seizure. This led to the film being reviewed by a customs officer. He then agreed to release the video after a short conversation: "You realize that this is a badly dubbed softcore movie from the 70s?" "Yep." "Uh—well, I hope you enjoy it." And we did indeed enjoy the scene with the binomial formula.

Infinite Worlds of H. G. Wells (2001), the hero is a seriously overweight mathematician. When he is talking about the circumference of the circle, with the relevant formula displayed on the blackboard, his pupils make fun of his "circumference." See also figure 18.3. In π (1998) the mathematical prodigy Max is looking for patterns in nature, in the stock market, and in π. While pondering π Max writes both formulas on a piece of newspaper. In *Red Planet Mars* (1952), a boy biting into a piece of pie has the idea of using the decimal expansion of π to communicate with aliens. His reasoning is that for the aliens to be able to build a radio dish they have to know the ratio of the diameter to the circumference, π.

The Volume of a Sphere

This can be found in the Pyecraft episode of *The Infinite Worlds of H. G. Wells* (2001) as well as in *St. Trinians* (2007). In both movies the volume formula is given incorrectly as πr^3. See figure 18.3.

21.18 Mathematical Games

21 (2008): A math professor organizes some of his students into a team of blackjack players. The Monty Hall puzzle has a large scene.

A Beautiful Mind (2001): A movie about the brilliant mathematician John Nash. In the extras on the DVD, there are some deleted scenes in which Nash is talking about the game of Hex, which the real John Nash invented.

Last Year at Marienbad (1961): A pretentious movie, which uses the mathematical game of Nim in an absurd manner. The starting position in this movie is always four rows of matchsticks. The rows are of sizes 1, 3, 5, and 7. A move is to take any number of matchsticks from one row only, removing them from the game. The game ends when the last object is taken. The player who takes the last object loses the game. With the particular starting position in *Marienbad*, the player who moves second can always force a win. A Nim-like game is also played by the detective and the mathematician murderer in the much more fun *Furuhata Ninzaburō: "Murder of a Mathematician"* (1995).

π (1998): The mathematical prodigy Max is looking for patterns in nature and in the stock market. He and his former supervisor are playing Go together, and Max uses the predictability of end games in this complex game to argue that there might be a pattern in similarly complex systems such as nature, the stock market, or the digits of π.

Teen Patti (2010): A weird Bollywood version of *21* (2008). Professor Subramaniam has some theory of randomness, and he and his students apply it to win at Teen Patti (basically three-card poker). The professor gives an awful and incorrect explanation of the Monty Hall problem (where the door to be opened is chosen randomly).

Movie Index

12 Angry Men (1957), 246, 251, 275
1984 (1984), 276
2010 (1984), 254
21 (2008), 248, 261, 268, 279
21 Grams (2003), 146
30 Virgins and Pythagoras (1977), 149, 152, 211, 252
3:19 (2008), 232
4D Man (1959), 175
The 4th Dimension (2006), 163, 171

The Abbott and Costello Show (1953), 119
Aces (2006), 275
The Adding Machine (1969), 241
A-Ducking They Did Go (1939), 275
After Midnight (2004), 261
Alien Hunter (2003), 179, 199, 249, 266, 272, 275
The Amateur (1981), 267
Amy & Isabelle (2001), 254
Andromeda (2000–2005), 226
Annapolis (2006), 208
Another Gay Movie (2006), 268
Antonia's Line (1995), 140, 191, 235, 239, 250, 260
Are You With It? (1948), 148, 248
The Arrival (1996), 200, 275
The A-Team (1983–1987), 272
Attack Girls' Swim Team Versus the Undead (2007), 219
Au Revoir Les Enfants (1987), 237, 254
The Avengers (1998), 170, 223, 249, 260
The Aviator (2004), 254

Back to the Future Part III (1990), 224, 258
Ball of Fire (1941), 227, 245, 276
The Bank (2001), 178, 185, 266, 272
Batman (1966), 211, 254
Battlefield Earth (2000), 211, 255, 277
Battle of the Worlds (1961), 185, 221, 269
A Beautiful Mind (2001), 21, 185, 234, 245, 250, 263, 267, 272, 279
Beautiful Ohio (2006), 249
Bedazzled (1967), 101, 273
Bedazzled (2000), 156, 197, 211, 237, 247, 252
Believers (2007), 243
Better Off Dead (1985), 143, 212, 237, 255
Bianca (1984), 241, 278
Big (1988), 246
The Big Bang Theory (2007–2011), 175, 221
Bill & Ted's Bogus Journey (1991),

182, 223, 272
Bingo (1991), 195, 224
Blackadder (1983–1989), 214
Blaise Pascal (1972), 232, 233
Bloodhounds of Broadway (1952), 204
Box of Moon Light (1996), 277
Brain Dead (1987), 177
Brave Archer 3 (1981), 194, 278
Breaking the Code (1996), 234, 247, 261, 267
Brutal (2007), 237, 241, 261
Buck Privates (1941), 198, 244
Buffy the Vampire Slayer (1997–2003), 212
Bullshot (1983), 225
Butterfly Dreaming (2008), 204

Cabiria (1914), 206, 231, 255
Caddyshack (1980), 255
Calabuch (1956), 277
Carry on Teacher (1959), 237
Cartesius (1974), 232, 233, 255
Casino Royale (2006), 241
Cast Away (2000), 246, 255, 278
Chaos (2005), 266
Check and Double Check (1930), 119
Chen Jingrun (2001), 97, 232, 233, 263
Chicago (2002), 275
City Homicide (2007–2011) 103, 241
Clan of the Cave Bear (1986), 235, 246
Class Action (1991), 241, 275
Cloak and Dagger (1946), 245, 269
Clue (1985), 217
Clueless (1995), 213
The Cocoanuts (1929), 247
Codename Icarus (1981), 133
The Colgate Comedy Hour (1952), 119
Colossus: The Forben Project (1970), 241, 269, 277
Conceiving Ada (1997), 232, 235, 249, 255

Contact (1997), 56, 202, 246, 250, 263, 267
Copenhagen (2002), 160, 246
The Core (2003), 202, 263
Cube (1997), 85, 163, 203, 240, 250, 263
Cube 2: Hypercube (2002), 96, 163, 235, 258, 267
Cutting Class (1989), 102, 198

Dangerous Sex Date (2001), 235, 269
The Da Vinci Code (2006), 246, 249, 261, 267
The Day of the Beast (1995), 267
Deadly Friend (1986), 268
Dead Poets Society (1989), 249
Dear Brigitte (1965), 239
Death of a Cyclist (1955), 242, 255
Death of a Neopolitan Mathematician (1992), 232, 243, 269
Deep Rising (1998), 220
Déjà Vu (2006), 171, 249, 258
Desk Set (1957), 219
Diabolique (1955), 197, 237
Diabolique (1996), 197, 242, 249 255
Dick Tracy (1990), 214, 247
Die Hard: With a Vengeance (1995), 102, 109, 220, 247, 250
Doctor Who (1963–1989, 2005–) 204, 262, 263
Donald in Mathmagic Land (1959), 71, 189, 201, 234, 250, 255
The Dot and the Line (1965), 255
Drawing Down the Moon (1997), 266
Dreams in the Witchhouse (2005), 255
Drowning by Numbers (1988), 252 272
Duck Soup (1933), 247
Dumbo (1941), 84

Der Elefant (2004–2006), 242
Eleventh Hour (2008–2009), 249, 261
The Englishman Who Went Up a Hill

and Came Down a Mountain (1995), 255

Enigma (2001), 146, 234, 268

The Enigma of Kaspar Hauser (1974), 101, 273

Entrapment (1999), 204

Eureka (2006–), 222

Eustice Solves a Problem (2004), 239

Évariste Galois (1965), 232

Event Horizon (1997), 171, 247, 258

Everybody Loves Raymond (1996–2005), 222

Evil Roy Slade (1972), 214, 244

Fahrenheit 451 (1966), 277

Family Guy (1999–), 212, 220, 277

The Favor (1994), 209, 266

Fermat's Last Tango (2001), 157, 211, 232, 234, 235, 253

Fermat's Room (2007), 97, 264, 274

First Circle (1992), 268

FlashForward (2009–2010), 199

Flash Gordon (1980), 224, 264

Flatland (1965), 175, 258

Flatlandia (1982), 175, 259

Flatland: The Film (2007), 175, 259

Flatland: The Movie (2007), 175, 259

The Flip Wilson Show (1970–1974), 119

Flubber (1997), 221, 249

Forbidden Planet (1956), 179, 267, 272

Force of Evil (1948), 277

Freaks and Geeks (2000), 205

Freaky Friday (2003), 218, 246, 247

Freedom Man (1989), 232, 246

Furuhata Ninzaburõ (1994–2006), 42, 242, 253, 279

Futurama (1999–2011), 97, 133, 163, 170, 220, 264, 269

The Gambler (1974), 276

George of the Jungle (1997), 246

The Giant Claw (1957), 62, 235

Good Will Hunting (1997), 3, 23, 146, 239, 246, 249

Graveyard Disturbance (1987), 170, 223, 260

Gulliver's Travels (1996), 232

Hands of a Murderer (1990), 268

Harold & Kumar Escape From Guantanamo Bay (2008), 227

Help (2005), 253

A Hill on the Dark Side of the Moon (1983), 234, 269

High School Musical (2006), 264

Homage (1995), 242

Hotel Hilbert (1996), 184, 272

How Green Was My Valley (1941), 219

How Not to Live Your Life (2007–2011), 18

The Hudsucker Proxy (1994), 248, 256

Ice Princess (2005), 249

The Ice Storm (1997), 145, 250, 256

Idiocracy (2006), 219

Il Posto (1961), 205

Imperativ (1982), 243

Infinitely Near (1999), 6

The Infinite Worlds of H. G. Wells (2001), 202, 237, 256, 264, 275

Infinity (1996), 182, 260, 272

An Innocent Love (1982), 270

Insignificance (1985), 175, 222, 259, 260

Inspector Lewis (2006), 97, 235, 242, 264

In the Navy (1941), 115, 212, 244

I.Q. (1994), 102, 180, 225, 232, 235, 248, 274

It's My Turn (1980), 133, 236, 250, 260, 264

I Went Down (1997), 219

Izo (2004), 260

Jam Films 2 (2004), 274

Jason X (2001), 226, 275

Jimtown Speakeasy (1928), 119
Julie Johnson (2001), 236, 244, 266
Jurassic Park (1993), 246, 266
Jurassic Park II (The Lost World)
    (1997), 246, 266
Jurassic Park III (2001), 266

The Karate Kid, Part II (1986), 204
The Keeper: The Legend of Omar
    Khayyam (2005), 233, 256
Kelly's Heroes (1970), 248
The King's Thief (1955), 234
Knocked Up (2007), 216
Kolchak (1974–1975), 267

Labyrinth (1986), 101, 223, 244, 245,
    274
Lambada (1990), 83, 195, 223, 237,
    256, 277
The Last Casino (2004), 256, 262
The Last Enemy (2008), 270
Las Vegas Shakedown (1955), 236,
    275
Las Vegas Weekend (1986), 184, 275
Late Bloomers (1996), 153
Law and Order: Criminal Intent
    (2001–), 104
Life is Beautiful (1997), 217
Like Mike (2002), 256
Little Big League (1994), 196, 220
Little Giant (1946), 119
Little Man Tate (1991), 83, 191, 216,
    238, 239, 246, 256
Loser Takes All! (2003), 276
Lost Souls (2000), 268
Love and Death (1975), 213, 244, 256
Love, Math and Sex (1997), 239, 253,
    260, 277
Lucky Pierre (1974), 144, 223, 238

Ma and Pa Kettle Back on the Farm
    (1951), 119
Madame Curie (1943), 256
Magic Town (1947), 249, 276
Magik and Rose (1999), 222, 266

Malcolm X (1992), 249
A Man Called Intrepid (1979), 268
The Man From Planet X (1951), 267
The Man Without a Face (1993), 153,
    206, 238, 246, 253, 256
Marius (1931), 216
Matemática Zero, Amor Dez (1958),
    157
Matilda (1996), 240
The Matrix Revolutions (2003), 170,
    223, 260
Meet Corliss Archer (1954), 205
Mean Girls (2004), 193, 238, 247, 270
Meet Dave (2008), 145
Mercury Rising (1998), 240
Merry Andrew (1958), 148, 152, 210,
    238, 247, 253, 257
The Mirror Has Two Faces (1958),
    130
The Mirror Has Two Faces (1996),
    121, 145, 183, 203, 244, 249, 250,
    264, 271, 273
Moebius (1996), 184, 260, 273
Monty Python and the Holy Grail
    (1975), 224
Monty Python Live at the Hollywood
    Bowl (1982), 221, 232, 233
Mozart and the Whale (2005), 215,
    240, 264
Mr. Bean (1990–1995), 214, 244
Mr. Magorium's Wonder Emporium
    (2007), 261
Münchhausen (1943), 278
The Music of Chance (1993), 203, 264
My Little Chickadee (1940), 226, 276
My Stepmother is an Alien (1988),
    227, 244
My Teacher's Wife (1995), 236, 238,
    244
Never Been Kissed (1999), 201, 244,
    262

Newton: A Tale of Two Isaacs (1997),
    234
No Highway in the Sky (1951), 97,

249

Northern Exposure (1992), 262

The Number 23 (2007), 65, 244, 252

Ocean's Eleven (2001), 244, 251

October Sky (1999), 238

Oh God! Book II (1980), 227, 244

Omar Khayyam (1957), 233, 249, 257

The Outer Limits (1963–1965), 259

Outside Providence (1999), 205

The Oxford Murders (2008), 242, 247, 250, 253

Paradisio (1961), 209

Parallel Life (2010), 243

Patch Adams (1998), 249

Peter the Great (1986), 234

The Phantom Tollbooth (1970), 183, 273, 274

Phase IV (1974), 257, 267

π (Pi) (1998), 53, 186, 201, 261, 262, 273, 279

Picnic at Hanging Rock (1975), 257

Pinky and the Brain (1995–2001), 260

The Pirates of Penzance (1983), 277

Pleasantville (1998), 170, 185, 223, 250, 260

Presumed Innocent (1990), 236, 242

The Prince and the Pauper (1990), 84, 238, 257

The Professor and His Beloved Equation (2006), 277

Proof (2005), 160, 224, 236, 247, 248, 264

Prophecy (1995), 249

Prospero's Books (1991), 257

A Pure Formality (1994), 147, 246, 257

Quality Street (1937), 218

Quest of the Delta Knights (1993), 232, 234

Race for the Bomb (1987), 234

Rain Man (1988), 246

Raising Genius (2004), 240

Real Genius (1985), 240

Recess (1997–2001), 18

Red Planet (2000), 247

Red Planet Mars (1952), 56, 187, 226, 262, 267, 279

The Return of Sherlock Holmes (1986), 257

Ring (1998), 214

The Road to Hong Kong (1962), 213, 246, 247, 277

Rocky and Bullwinkle (1959–1961), 234

Rosario + Vampire (2008), 154

Rosencrantz & Guildenstern are Dead (1990), 226, 276

The Rules of Attraction (2002), 209, 276

Rushmore (1998), 238, 240, 258, 271

The Saragossa Manuscript (1965), 178, 190, 273

Schoolgirl Report Part 4 (1972), 278

The School of Rock (2003), 227, 238, 244

Sebastian (1968), 268

Sekret Enigmy (1979), 268

Serial Mom (1994), 243

The Seven-Per-Cent Solution (1976), 243, 271

Sex in the City (1998–2004), 248, 264, 276

She Wrote the Book (1946), 236, 271

Shrieker (1998), 173, 204, 236, 259

The Siege of Syracuse (1960), 221, 231, 258

The Simpsons (1989–) 151, 156, 161, 223, 227, 253, 259, 262, 276

Si Versailles M'Était Conté (1954), 231

Six Feet Under (2001–2005), 238

Sky Captain and the World of Tomorrow (2004), 248

Small Time Crooks (2000), 216, 244

Smilla's Sense of Snow (1997), 147, 236

Sneakers (1992), 250, 264, 268

Solid Geometry (2002), 247, 258

Son of Sinbad (1955), 233, 248

Southpark (1997–), 220

Spaceways (1953), 236

The Spanish Prisoner (1997), 140, 261

Sphere (1998), 247, 258, 267

Spider-Man 2 (2004), 247

Stand and Deliver (1988), 41, 213, 238, 248, 271

Stand-In (1937), 149, 258

Stargate (1994), 206, 249

Stargate: Atlantis (2004–2009), 203

Star Trek (1966–2005) 155, 199, 234, 240, 248, 249, 253, 258, 262, 273

The Story of Mankind (1957), 247

Strangers on a Train (1951), 271

Straw Dogs (1971), 214, 243, 246

St. Trinians (2007), 202, 279

Succubus (1968), 209, 258

Super Mario Bros. (1993), 205

Supernova (2000), 170, 249, 259

Suspect X (2008), 243

Taken (2002), 261

Team America: World Police (2004), 212

Teen Patti (2010), 247, 279

Teresa's Tattoo (1994), 143, 223, 236

That's Adequate (1989), 221

Threshold (2005–2006), 133, 204, 259, 265, 266, 276

Thunderbirds (1965–1966), 198, 226

Titanic (1997), 250

Tom & Viv (1994), 102, 194, 234

Torchy Gets Her Man (1938), 226

Torn Curtain (1966), 248, 263, 272

To Sir, with Love (1967), 248

Twilight (2008), 263

The Twilight Zone (1959–1964, 1985–1989), 204, 224, 246

Twin Peaks (1990–1991), 218

Unabomber: the True Story (1996), 233

Up in the Air (2009), 244

Village of the Damned (1960), 240

The Virgin Suicides (1999), 201, 238, 263

Waterboys (2001), 258

Willy Wonka & the Chocolate Factory (1971), 239, 276

Wish Upon a Star (1996), 197, 239

Wittgenstein (1993), 234

The Wizard of Oz (1939), 151, 210, 253

Wonder Woman (1975–1979), 221

The Wonder Years (1988–1993), 239

The World is Not Enough (1999), 204

Yes Sir, Mr. Bones (1951), 119

Young Einstein (1988), 240

Zhukovsky (1950), 235